Stefanie Grunwald

Muscular Dystrophy Duchenne modifying genes and pathways

Stefanie Grunwald

Muscular Dystrophy Duchenne modifying genes and pathways

An interdisciplinary approach of molecular biology and systems biology

Südwestdeutscher Verlag für Hochschulschriften

Impressum/Imprint (nur für Deutschland/ only for Germany)
Bibliografische Information der Deutschen Nationalbibliothek: Die Deutsche Nationalbibliothek verzeichnet diese Publikation in der Deutschen Nationalbibliografie; detaillierte bibliografische Daten sind im Internet über http://dnb.d-nb.de abrufbar.
 Alle in diesem Buch genannten Marken und Produktnamen unterliegen warenzeichen-, marken- oder patentrechtlichem Schutz bzw. sind Warenzeichen oder eingetragene Warenzeichen der jeweiligen Inhaber. Die Wiedergabe von Marken, Produktnamen, Gebrauchsnamen, Handelsnamen, Warenbezeichnungen u.s.w. in diesem Werk berechtigt auch ohne besondere Kennzeichnung nicht zu der Annahme, dass solche Namen im Sinne der Warenzeichen- und Markenschutzgesetzgebung als frei zu betrachten wären und daher von jedermann benutzt werden dürften.

Verlag: Südwestdeutscher Verlag für Hochschulschriften GmbH & Co. KG
Dudweiler Landstr. 99, 66123 Saarbrücken, Deutschland
Telefon +49 681 37 20 271-1, Telefax +49 681 37 20 271-0
Email: info@svh-verlag.de
Zugl.: Berlin, Humboldt-University, 2009

Herstellung in Deutschland:
Schaltungsdienst Lange o.H.G., Berlin
Books on Demand GmbH, Norderstedt
Reha GmbH, Saarbrücken
Amazon Distribution GmbH, Leipzig
ISBN: 978-3-8381-2080-5

Imprint (only for USA, GB)
Bibliographic information published by the Deutsche Nationalbibliothek: The Deutsche Nationalbibliothek lists this publication in the Deutsche Nationalbibliografie; detailed bibliographic data are available in the Internet at http://dnb.d-nb.de.
 Any brand names and product names mentioned in this book are subject to trademark, brand or patent protection and are trademarks or registered trademarks of their respective holders. The use of brand names, product names, common names, trade names, product descriptions etc. even without a particular marking in this works is in no way to be construed to mean that such names may be regarded as unrestricted in respect of trademark and brand protection legislation and could thus be used by anyone.

Publisher: Südwestdeutscher Verlag für Hochschulschriften GmbH & Co. KG
Dudweiler Landstr. 99, 66123 Saarbrücken, Germany
Phone +49 681 37 20 271-1, Fax +49 681 37 20 271-0
Email: info@svh-verlag.de

Printed in the U.S.A.
Printed in the U.K. by (see last page)
ISBN: 978-3-8381-2080-5

Copyright © 2010 by the author and Südwestdeutscher Verlag für Hochschulschriften GmbH & Co. KG and licensors
All rights reserved. Saarbrücken 2010

*For my family and
my unborn child*

Abstract

Background and aim: DMD is the most common muscular dystrophy in childhood and incurable to date. It is caused by the absence of dystrophin, what influences several signal transduction pathways. The thesis is interested in the investigation and modulation of signal transduction pathways that may compensate the lack of dystrophin as an alternative therapy strategy.

Experimental strategy: To study dystrophin downstream pathways the mRNA expression of DMD patients and two DMD siblings with an intra-familially different course of DMD were analysed in muscle tissue. On the basis of these expression data a Petri net was first developed implicating signal transduction pathways and dystrophin downstream cascades. Invariant (INA) and theoretical knockout (Mauritius Maps) analyses were applied for studying network integrity and behaviour. Both methods provide information about the most relevant part of the network. In this part modulation of protein activity and of gene expression using siRNA, vector-DNA, and chemical substances were performed on human SkMCs. Subsequently, the cells were studied by proliferation and vitality tests as well as expression analyses at mRNA and protein level.

Results: RAP2B and CSNK1A1 were differently expressed in two DMD siblings, and first are part of a signal transduction pathway implicating dystrophin downstream processes. The central point of this pathway is the de- and activation of the transcription factor NFATc. Its target genes are, among others, the negative proliferation factor p21, the Dystrophin homologue UTRN, and the differentiation factor MYF5. Consequently, an increase in UTRN implicates an undesirably reduced myoblast proliferation rate. Latter was found in DMD patients and was target for further studies. But, siRNA and vector DNA experiments showed that NFATc is not the decisive factor for the target genes. Deflazacort and cyclosporin A are known to influence the activation of NFATc. The results first showed that both substances do induce myoblast proliferation. The use of deflazacort in combination with cyclosporin A resulted in an increase of UTRN expression.

Conclusion: The modulation of proliferation and UTRN-expression independently of each other is possible. According to the basic idea of this study, a new therapeutic strategy becomes apparent, which considers dystrophin downstream processes.

Acknowledgement

This thesis was performed at the Humboldt-University and the University of Applied Sciences in Berlin.

My utmost gratitude goes to Prof. Dr. Speer for the great counselling and support, for lending me an ear at anytime, and for the confidence having placed in me to work on this fascinating and multifarious project. Furthermore, I would like to express my sincere appreciation to Prof. Dr. Koch for introducing in the world of Petri nets and mentoring of my thesis.

My thanks and appreciation goes to my advisor Prof. Dr. Herzel at the Humboldt-University for his encouragement, and who assisted me with this dissertation through his thoughtful advices.

The work was supported by an operating grant from the German Society of Muscle Diseased Persons and a studentship from the "Hypatia Programm" of the Technical University of Applied Sciences Berlin, Germany, and the "Berliner Programm zur Förderung der Chancengleichheit für Frauen in Forschung und Lehre", Humboldt-University Berlin.

Table of contents

Abstract 3

Acknowledgement 4

Abbreviations 9

1   Introduction 11
  1.1 Duchenne muscular dystrophy – general aspects 11
  1.2 Molecular basis of Duchenne and Becker muscular dystrophy 11
  1.3 Therapeutic strategies 14
      1.3.1 Substitution and correction of dystrophin 16
      1.3.2 Elevation of utrophin 20
      1.3.3 Dystrophin downstream strategies 23
  1.4 Interest of systems biology 27
      1.4.1 Theoretical methods in systems biology 29
      1.4.2 Petri net basics 31
      1.4.3 Invariant analysis 33
      1.4.4 Maximal common transition set 35
      1.4.5 Cluster analysis 35
      1.4.6 Mauritius map 36
      1.4.7 Application of Petri nets to biochemical systems 37

2   Conceptual formulation 39

3   Material and Methods 40
  3.1 Material 40
      3.1.1 Chemicals 40
      3.1.2 Buffers and solutions 42
      3.1.3 Media for cell culture 43
      3.1.4 Oligonucleotides 44
      3.1.5 Kits 45
      3.1.6 Molecular weight marker and plasmid vector DNA 46
      3.1.7 Enzymes 46
      3.1.8 Antibodies 47
      3.1.9 Transfection reagents and systems 47
      3.1.10 Equipment 48

	3.1.11	Consumable supplies 49
3.2	Software 50	
3.3	Methods 51	
	3.3.1	Flow diagram 51
	3.3.2	Origination of muscle tissue and human skeletal muscle cells 52
	3.3.3	Culturing of primary human skeletal muscle cells (myoblasts) 54
	3.3.4	Transfection 56
	3.3.5	Treatment of primary human myoblasts with chemical substances 63
	3.3.6	Cell Proliferation and vitality tests 66
	3.3.7	Isolation of DNA and RNA 67
	3.3.8	Reverse transcription of DNA-free RNA 70
	3.3.9	PCR 71
	3.3.10	Sequencing 74
	3.3.11	Protein analysis 75
3.4	Bioinformatics tools 78	
4	Results 79	
4.1	Signal transduction pathways connecting genes which were found differentially expressed in transcriptoms of selected DMD patients 79	
4.2	Verification and extension of mRNA expression data 83	
	4.2.1	Muscle tissues originating from Duchenne muscular dystrophy patients and controls 83
	4.2.2	Expression analysis of selected genes at mRNA level 84
4.3	Systems biology 86	
	4.3.1	The Petri net model 86
	4.3.2	Structural analysis of the model 90
4.4	Modulation of gene expression of CSNK1A1 by plasmid vector cDNA as well as siRNA and modification of CSNK1A1 and calcineurin activities using chemical substances 103	
	4.4.1	Characterisation of cell cultures originating from DMD patients and controls 104
	4.4.2	Applications governing CSNK1A1 activity and expression 104
	4.4.3	Up-regulation of CSNK1A1 using plasmid vector cDNA 106

		4.4.4 Modulation of calcineurin signalling 115

5 Discussion 126

   5.1 Sub-net connecting dystrophin with p21 and UTRNA via phosphorylation of NFATc through CSNK1A1 and JNK1 129

        5.1.1 Petri net analyses of the dystrophin downstream sub-net 129

        5.1.2 RNA expression studies of selected components 131

        5.1.3 Experimental modulation of CSNK1A1 expression 133

   5.2 Subnet connecting de-phosphorylation of NFATc via RAP2B-triggered calcineurin activation 135

        5.2.1 Petri net analyses of the RAP2B-NFATc sub-net 135

        5.2.2 RNA expression studies of selected components of the RAP2B downstream cascade 137

        5.2.3 Modulation of calcineurin activity using pharmacological interventions 139

   5.3 Future directions 142

6 Summary 144

7 Zusammenfassung 146

8 References 149

9 Appendices 165

   9.1 Glossary 165

   9.2 Publication list 168

Abbreviations

ABI	Applied Biosystems
aDG	α-dystrobrevin
AON	antisense oligonucleotides
APS	ammonium persulfate
BCA	bicinchoninic acid
bDG	β-dystrobrevin
BMD	Becker muscular dystrophy
BrdU	5-bromo-2'-deoxyuridine
BSA	Bovine serum albumin
CARP	cardiac ankyrin repeat protein
CBFB	core binding factor beta
CDK	cyclin-dependent kinase
CMV	cytomegalovirus promoter
Cryo-SFM	cryo-serum free medium
CsA	cyclosporin A
CSNK1A1	casein kinase 1 alpha 1
CPN	coloured Petri Nets
C_T	cycle threshold
CTI	covered by t-invariants
DCTN3	dynactin 3 light chain
ddNTPs	dideoxynucleotide
DEPC	diethyl dicarbonate (Diethylpyrocarbonate)
DGC	dystrophin-glycoprotein complex
DIF	desmin intermediate filaments
DMD	duchenne muscular dystrophy
DMEM	dulbecco's Modified Eagle's Medium
DMSO	dimethyl sulfoxide
DRP	dystrophin-related proteins
DTT	dithiothreitol
ECL	enhanced chemiluminescence
EDTA	ethylene diamine tetraacetic acid
eGFP	enhanced GFP
ER	endoplasmatic reticulum
FCS	fetal calf serum
FRET	fluorescence resonance energy transfer
GAPDH	glyceraldehyde 3-phosphate dehydrogenase
GFP	green fluorescent protein
Grb2	growth factor receptor-bound protein 2
IC261	3-[(2,4,6-trimethoxyphenyl)methylidenyl]-indolin-2-one
IP3	Inositoltrisphosphat
IRES	internal ribosome entry site
LB	lysogeny broth (Luria-Bertani broth)

ABBREVIATIONS

m	marking
MCA	metabolic control analysis
MCT-set	maximal common transition set
MGB	minor groove binder
MLCf1	Myosin light chain 1
MTCC	Muscle Tissue Culture Collection
NFATc	nuclear factor of activated T-cells
NHDF	human dermal fibroblast nucleofector kit
nNOS	neuronal NO synthase
NO	nitric oxide
ODE	Ordinary Differential Equation
ODN	oligodesoxynucleotide -mediated gene correction
OMIM	Online Mendelian Inheritance in Man
P	place
p21	cyclin-dependent kinase inhibitor 1A
PBS	phosphate buffered saline
PCR	polymerase chain reaction
p-invariant	place invariant
p/t net	place/ transition net
PIP2	phosphatidylinositol bisphosphate
POD	peroxidase
RAP2B	ras related protein 2 beta
RIPA	radio-immuno-precipitation assay
S	syncoilin
SCID-Xl	X-linked severe combined immune deficiency
SDS-PAGE	sodium dodecyl sulfate poly-acryl-amide gel electrophoresis
SG	sarcogylan complex
siRNA	short interfering RNA
SkMCs	skeletal muscle cells
Spn	sarcospan
Syn	syntrophins
T	transition
TAE	Tris-acetate-EDTA
TBS	Tris-buffered saline
TBS-T	Tris-buffered saline with Tween 20
TFB	transforming buffer
t-invariant	transition invariant
TNS	trypsin neutralisation solution
UPGMA	unweighted pair group method with arithmetic mean
UTRNA	utrophin A
WST-8	2-(2-methoxy-4-nitrophenyl)-3-(4-nitrophenyl)-5-(2,4-disulfophenyl)-2H-tetrazolium, monosodium salt
XLDC	X-linked dilated cardiomyopathy

1 Introduction

Molecular biology

1.1 Duchenne muscular dystrophy – general aspects

Duchenne muscular dystrophy (DMD) is an X-linked recessive disease with a birth prevalence of 1 in 3500 live born males and usually affects boys. The first clinical symptoms appear at the age of 2 to 4 years.
DMD patients often show a delay in milestones of the development in early childhood including delays in sitting and standing independently followed by developing of gait abnormalities such as difficulties in climbing stairs and waddling walking. Some patients also show mental retardation. At the age of 9 to 12 years the patients are wheelchair-dependent. In their early teens, the patients suffer from cardiomyopathy and/or respiratory complications. At the end of the second or at the beginning of the third decade, DMD patients die due to cardiac insufficiency or respiratory problems.
Becker muscular dystrophy (BMD) is characterised by a later onset and a milder progression of the disease with a birth prevalence of 1 in 18500 live born males. The patients become wheelchair-dependent in their 3rd of 4th decade. The age of death varies from 40 to 60 years, and heart failure is the common cause of morbidity (Pongratz *et al.*, 1990).

1.2 Molecular basis of Duchenne and Becker muscular dystrophy

DMD is a consequence of the loss of the protein dystrophin. BMD patients still posses a truncated form of dystrophin. In 1986, the gene *dystrophin* was identified by positional cloning at the location Xp21 (Monaco *et al.*, 1986). It spans 2.4 Mb, consists of 79 exons, and results preferentially in a 14kb transcript, which encodes the 427kD dystrophin protein. But, the *dystrophin* gene also causes different mRNA-transcripts encoding various dystrophin isoforms. The expression of these isoforms is regulated by 7 tissue-specific promoters as well as alternative splicing of the pre-mRNA. The length of the known dystrophin isoforms and their tissue-specific expression state are summarised in figure 1.

INTRODUCTION

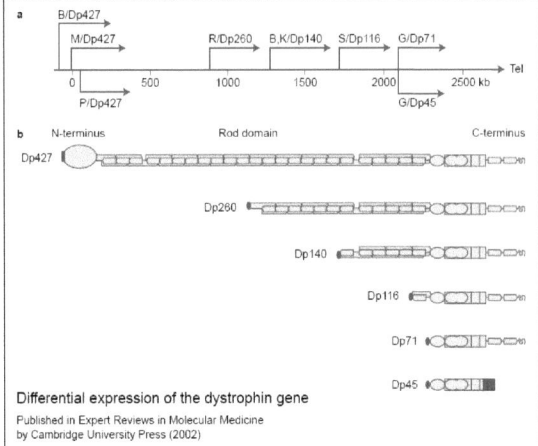

figure 1: Differential expression of the dystrophin gene (Wells *et al.*, 2002)
  a) The position of the different promoters in the dystrophin gene (B, brain; M, muscle; P, Purkinje; R, retina; K, kidney; S, Schwann cells; G, general)
  b) The 6 isoforms of dystrophin and their domains. Dp427 include the N-terminal actin-binding domain. The other isoforms contain some triple-helical coiled-coil repeats and the C-terminus (repeats = parallel rod; hinges = notch)

The most frequent isoform of dystrophin is expressed full length with 3685 amino acids in the skeletal and cardiac muscle and located in the sub-sarcolemmal region of the muscle fibre in humans (Hoffman *et al.*, 1987). It consists of 4 distinct domains (Koenig *et al.*, 1988) as shown in figure 2.

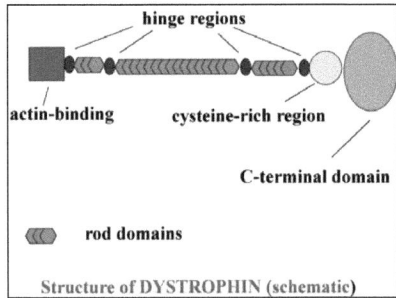

figure 2: The domain composition of the full-length protein dystrophin. The N-terminal domain mediates binding of γ-actin binding filaments. The subsequent central rod domain contains 4 hinge regions followed by a cystein-rich domain responsible for interaction with integral membrane proteins. A unique C-terminus completes the protein dystrophin.
[http://www-biology.ucsd.edu/classes/bimm110.SP07/images/dystrophin.jpg]

The domain at the N-terminal end presents homology with actin-binding proteins such as α-actinin and ß-spectrin. It is responsible for binding the γ-actin filaments and comprises between 232 and 240 amino acids. The following central rod domain is the largest domain of the protein and consists of 24 triple helical spectrin-like repeats interspersed with 4 putative hinge domains. The end of the rod domain is associated with the third domain, a cysteine-rich domain with 2 EF-hand motifs. This region interacts with integral membrane proteins, including sarcoglycan and dystroglycan, members of a complex known as dystrophin-glycoprotein complex (DGC) (see section 1.3.3.1). This domain is thought to provide the essential anchor of dystrophin to the cell membrane and consequently to the sarcolemma. The carboxy terminus is unique to dystrophin with the exception that 2 other proteins show homology to this region: utrophin (UTRN) and dystrobrevin. Dystrophin consists of an α-helical structure, which interacts with syntrophin, and dystrobrevin. Syntrophin, dystrobrevin, sarcoglycan and dystroglycan form the DGC. The first and the last part of the N-terminal and the C-terminal domains are assumed not to be essential for the development of the DGC (Ervasti, 2007).

Mutations in dystrophin comprise deletions, insertions, and point mutations. The mutations can result in a premature stop of the translation or a modified structure due to a different amino acid sequence (Roberts *et al.*, 1994; Legardinier *et al.*, 2009) that lead to a functional impairment of the protein. Approximately 30% of the mutations occur spontaneously. The most common mutations in dystrophin are deletions of single or multiple exons, which applies to 65% of the DMD patients. Point mutations are found in around 30% of the cases, resulting in a shift of the open reading frame or directly in an early stop codon. The remaining patients are found to have insertions.

At first, diagnostics include a blood test to measure serum creatine kinase that is increased at least 10fold as a consequence of a myogenic degradation. An electromyography is used to test the electrical activity of the musculature, which offers the distinction between a myositis, a dystrophy and a disturbed neural maintenance. Detection of mutations by PCR is the main point in molecular diagnostics. Routinely, up to 85% of the mutations can be detected by including deletions and duplications. Point mutations are also identifiable via high-throughput denaturing HPLC followed by direct sequencing. In spite of this, diagnostic procedures are usually not applied because of the size of dystrophin gene on the one hand and the rareness of the cases on the other (Trimarco *et al.*, 2008). In cases of undetected mutations, a muscle tissue sample is taken from the patients to verify the absence of dystrophin at protein level using western blot analysis or immunohistochemical analyses. In BMD patients, the protein level of dystrophin is changed but still detectable. Muscle tissue revealed from DMD patients also shows abnormal variation in the diameter of the muscle fibres with atrophic, hypertrophic and necrotic fibres and an increase of endomysial connective and fat tissue (Speer and Oexle, 2000).

1.3 Therapeutic strategies

Whereas diagnostics can be considered satisfactory, no causal and/or efficient therapy is currently available for DMD. The patients are presently treated with corticosteroids such as prednison or deflazacort, physical therapy, and orthopaedic equipment as well as orthopaedic surgery and respiratory assistance.
For most of the laboratory investigation of therapeutic strategies, cell culture systems such as primary human myoblasts or DMD model organisms such as the mdx mouse or the DMD-dog are used. Both, the original mdx mouse and the DMD-dog are species with a naturally occurring lack of dystrophin. Additional transgenic mouse models are also designed with different introduced mutations in the *dystrophin* gene. Whereas the DMD-dog, often a golden retriever, presents a phenotype like that found in humans, mdx mice are clinically only marginally affected, which leads to the assumption that the mdx mouse by-passes the defect by means of an elusive signalling pathway (Vainzof *et al.*, 2008).
There are many approaches being discussed to reduce progression of the disease or to restore muscle function. They can be classified into gene-therapeutic, cell-based and pharmacological strategies and include the substitution and correction of dystrophin, up-regulation of utrophin A (UTRNA) and dystrophin downstream signalling cascades (Manzur *et al.*, 2008b). An overview of therapy strategies and medical treatment is given in table 1 (Sakamoto *et al.*, 2002; Zhou *et al.*, 2006). Selected aspects itemised in table 1 are presented in more detail in section 1.3.1 and 1.3.2.

table 1: Therapeutic strategies and current pharmacological treatment (modified after Sakamoto et al. and Zhou et al. (Sakamoto et al., 2002; Zhou et al., 2006)). As a rule, publications represent review article. If available, references that discuss or refer to appropriate clinical studies are also included. The list does not claim to be complete.

				Clinical trials
Molecular therapy	Substitution of the dystrophin gene	Plasmid DNA (Rando, 2007) Adeno-viral vector (Odom et al., 2007) Adeno-associated vector (Odom et al., 2007) Lenti-viral vector (Odom et al., 2007)		(Fardeau et al., 2005; Duan, 2008)
	Correction of the mutation	RNA level	Viral-directed exon skipping (Bertoni, 2008) Antisense oligonucleotides producing skipping of mutated exons (Rando, 2007)	(Bertoni, 2008; Muntoni et al., 2008)
		DNA level	Oligodesoxynucleotide-mediated gene correction (ODN) (Bertoni, 2005)	
	Up-regulation of dystrophin-related proteins (DRP)	Plasmid DNA bearing DRP (Miura and Jasmin, 2006) Plasmid DNA containing transcription factors of the DRP (Miura and Jasmin, 2006)		
Cell-based strategies	Stem cell therapy (Grounds and Davies, 2007)	Satellite cells (Zhou et al., 2006) Adipose, muscle- or bone-marrow-derived side population cells (Vieira et al., 2008) Embryoid body-derived cells (Shao et al., 2009)		
	Myoblasts (Grounds and Davies, 2007)	Myoblast transfer		(Skuk et al., 2007)
Pharmacological treatment	UTRN	up-regulation of UTRN (Miura and Jasmin, 2006)		
	Bypassing stop codons (Schmitz and Famulok, 2007)	Gentamicin PTC124		(Politano et al., 2003) (Hirawat et al., 2007)
	Muscle repair (Zhou et al., 2006)	IGF-1 Inhibtion of myostatin		
	Protease inhibitors (Zhou et al., 2006)	Calpain inhibition Serin protease inhibition Proteosome inhibition		
	Anti-inflammatory drugs	Cyclosporin A (Sharma et al., 1993) Prednison/ Deflazacort (St-Pierre et al., 2004)		(Sharma et al., 1993) (Korinthenberg, 2008) (Houde et al., 2008; Manzur et al., 2008a; Mavrogeni et al., 2009)
	Anti-oxidant agent	Idebenone (Buyse et al., 2009)		
	Calcium channel blockers	Diltiazem/Verapamil (Matsumura et al., 2009)		(Phillips and Quinlivan, 2008)

1.3.1 Substitution and correction of dystrophin

1.3.1.1 Dystrophin substitution

DMD is a monogenic disease. The first attempt was to correct the defect by substitution of dystrophin. In early studies, the dystrophin gene was transferred into the mdx mouse, which led to a reverse of the disease symptoms. The simplest way to deliver the dystrophin gene is the intravenous or intramuscular injection of plasmid DNA containing the dystrophin gene (Fardeau *et al.*, 2005; Duan, 2008). Disadvantages of this method are an activation of antibody-mediated immune response because of its neo-antigen character for the cells, the short retention time, and the poor distribution of the plasmid DNA with a distance of a couple of centimetres to injection site. In France, a phase I clinical trial with DMD-patients involving plasmid DNA containing dystrophin confirmed these findings. Despite the disadvantages mentioned above, plasmid DNA approaches have the advantage of avoiding additional immune response as would be expected if viral vectors were used.

Gene delivery by using viral vectors is slightly more efficient than plasmid DNA. The substituted dystrophin gene may be able to integrate into chromosomal DNA and permanently correct the gene defect. Apart from the immune response, the viral vectors may trigger oncogenesis in consequence of their integration site, especially by the use of lenti-viral vectors. Currently, 3 different viral vectors are applied in studies: adeno-viral, adeno-associated-viral and lenti-viral vectors (Odom *et al.*, 2007).

By nature, adeno-viral vectors possess a cloning capacity of 8kb, which could only include a truncated form of dystrophin. The new generation of fully truncated adeno-viral vectors offer a cloning capacity of 36kb. This vector can carry the full length cDNA of dystrophin, but also presents reduced immune response because of the lack of viral proteins (Ohshima *et al.*, 2009). It was demonstrated that 52% of muscle fibres were dystrophin-positive after an infection with a fully truncated adeno-virus bearing 2 copies of the murine full-length dystrophin (Dudley *et al.*, 2004).

The adeno-associated viruses are advantageous because of the non randomised integration into the chromosomal locus 19q13, which reduces enormously the risk of oncogenesis and the inability for replication. The latter diminish the immune response. Furthermore, recombinant vectors do not include any viral genes, but still the capsid proteins may induce humoral immune response. The cloning capacity represents 5kb, which only allows a transfer of a micro-dystrophin. Another limitation is that adeno-associated viruses rarely integrate into chromosomes, and may consequently be lost after several cell cycles. A modified adeno-associated vector carrying a micro-dystrophin is used in a clinical Phase I/IIa study at Columbus Children's Hospital in Ohio, USA. It is injected into the biceps of DMD patients (Muntoni and Wells, 2007).

Lenti-viruses, in particular the HI-virus, are used for delivering of a 6kb mini-dystrophin in proliferating but also in differentiated cells. In contrast to adeno-viral vectors, lenti-viruses do not induce a cytotoxic T-cell lymphocyte response. Also, a self-inactivating lenti-viral vector was engineered to exclude safety concerns. The main disadvantage is the rather unspecific integration of the virus into chromosomal DNA, which might lead to oncogenesis (Romano, 2005). In France, a clinical trial on patients suffering from X-linked severe combined immune deficiency (SCID-Xl) led to the development of leukaemia in 2 of 11 patients after treatment with a retro-viral vector (Hacein-Bey-Abina *et al.*, 2003a; Hacein-Bey-Abina *et al.*, 2003b). The study of Schroeder et al. demonstrated regional hotspots for integration of the HI-virus, and also showed that active genes were preferential integration targets (Schroder *et al.*, 2002). Up to now, the integration site of lenti-viral vectors cannot be predicted with certainty, but attends to influence the integration target are currently being made. The use of a novel system based on integrase-deficient lenti-viral vectors allows efficient gene transfer without challenges of lenti-viral vectors integration (Philpott and Thrasher, 2007).

1.3.1.2 Dystrophin correction

The substitution of a correct dystrophin gene is possible, but complex due to the size of the gene and the difficulty of safe and efficient delivery of the vector. A correction of the dystrophin gene would be feasible at RNA and DNA level.

At the RNA level, short DNA sequences, called antisense oligonucleotides (AON), are able to recognise sequences at the mRNA level specifically resulting in different effects such as gene knockdown, gene up-regulation or shifting of the reading frame (Rando, 2007; Bertoni, 2008; Mitrpant *et al.*, 2009). Originally, AON are designed to knock down the expression of a target gene by binding complementary to the mRNA, which possibly blocks or inhibits protein expression. But, exon-specific AON can concurrently block splicing enhancers or sequence regulatory elements that control exon recognition induces skipping of the exon(s) that may carry an out-of-frame mutation. This restores coding reading frame leading to expression of an in-frame mRNA encoding for a shorter protein that may be still functional (Doran *et al.*, 2009). Thus a DMD phenotype could be transformed in BMD. Aartsma-Rus and van Ommen suggest that multi exon skipping of 20 exons in the central rod domain would theoretically be beneficial for at least 63-75% of all DMD patients (artsma-Rus and van Ommen, 2007; Beroud *et al.*, 2007; artsma-Rus *et al.*, 2009). However, in a more recent study the authors report on a multiexon 45-55 skipping that led to a minimal skip frequency comparable to controls (van Vliet *et al.*, 2008). The central rod domain is the largest domain of the protein, but does not contain the essential DGC or actin-binding elements. Certain BMD-like patients demonstrate the expression of an in-frame transcript encoding for a much shortened but extremely functional form of dystrophin.

In 1995, the pioneering work of the group of Takeshima et al. was able to modulate exon skipping of dystrophin exon 19 in an *in vitro* system. For that purpose, artificial dystrophin mRNA precursors were established containing exon 18, a truncated intron 18, and exon 19. An antisense 31-mer 2'-O-methyl ribonucleotide, complementary to the splicing recognition sequence of exon 19 inhibited splicing of wild type pre-mRNA in a dose- and time-dependent manner (Takeshima *et al.*, 1995). Since then, exon skipping is discussed as a new therapeutic approach for DMD (Muntoni and Wells, 2007; Bertoni, 2008). One of the main limitations of this approach is the half-life of skipped RNA and truncated protein as well as the continuous administration of AON to a large area and number of fibres. The most favourable candidates for a specific delivery of AON are viral vectors, but they are of course subject to the hurdles as described for dystrophin substitution. The safety of the AON itself has been described as fairly good from data received from human trials, but toxicities in terms of hybridisation have been reported (Muntoni and Wells, 2007). Recently, Neri et al. reported that dystrophin protein levels between 29% and 57% were found in families with X-linked dilated cardiomyopathy (XLDC) in a completed clinical trial phase I A. XLCD patients carry a mutation at the 5' end of the gene and typically suffer from myopathy restricted to the cardiac muscle, but possess dystrophin in the skeletal muscle. So it is assumed that a level of dystrophin of 30% of normal would be sufficient to avoid muscular dystrophy (Neri *et al.*, 2007). The efficiency of different chemical variants of AONs are currently being studied (Heemskerk *et al.*, 2009). However, since 2006, in the Netherlands a clinical Phase I/II study has been conducted on DMD patients using RNA-based therapeutic PRO051, and is still ongoing (van Deutekom and van Ommen, 2006; van Deutekom *et al.*, 2007). Current clinical trials using AON are summarised by Muntoni et al. (Muntoni *et al.*, 2008).

At the DNA-level, the oligodesoxynucleotide-mediated gene correction (ODN) is based on a designed mismatch between the targeting vector and the genome sequence of interest inducing a desired single base pair change through endogenous DNA repair (Bertoni, 2005;

Rando, 2007). Thus, DMD patients carrying a point mutation would benefit from this therapeutic strategy, which applies to approximately 10-15% of all patients (Muntoni et al., 2003). In several reports, a correction of the dystrophin gene was published using ODN restoring of dystrophin expression. The main disadvantages of this technique are the low efficiency of delivering and the effect of ODN. However, the greatest advantage of ODN-mediated gene correction is the permanent correction of the gene defect and consequently constitutes the most efficient therapeutic strategy for DMD patients.

1.3.1.3 Bypassing of stop-codons

The idea of a read-through strategy is based on the fact that some antibiotics, such as aminoglycosides, are able to repress stop codons. The medication of patients suffering from DMD due to a stop codon with such an antibiotic may lead to read-through of the stop-codon resulting in a functional protein (Schmitz and Famulok, 2007). A study using gentamicin in mdx mice was able to demonstrate an increased expression of dystrophin of 20% and a reconstitution of the DGC at the sarcolemma. Subsequent clinical trials with DMD and BMD patients (each with 2) treated with gentamicin did not result in any appreciable detection of full-length dystrophin (Politano et al., 2003). These disappointing results in addition to the eventual toxicity of a permanent intake of antibiotics initiated the development of new strategies. The biopharmaceutical company *PTC Therapeutic* discovered an orally administered, small-molecule drug called PTC124. PTC124 targets post-transcriptional control processes allowing ribosomes to bypass the nonsense mutation in mRNA. In mdx mice, its efficiency to initiate the production of full-length dystrophin has been determined to be up to 20-25% that of the control mice muscles (Welch et al., 2007). A change concerning a read-through of physiological stop codons could not be detected since mRNA typically carries multiple stop codons at the 3´ end (Muntoni and Wells, 2007). The Phase I study has indicated that PTC124 is well tolerated with little adverse effects at the highest dosage (Hirawat et al., 2007). Currently, PCT124 is in Phase II clinical trials with promising preliminary results. About 50% demonstrated an increase in the immunostaining expression of dystrophin. Analysis of the data is still ongoing. However, the efficiency of PTC124 is limited due to its varying ability to enhance read-through of different stop sequences. *In vitro* experiments show suppression of UGA>>UAA~UAG, but most readily if the specific sequence is UGAC. Since fewer than 10% of all DMD patients carry a nonsense mutation, only a few percent may benefit from PTC124 (Aurino and Nigro, 2006; Wilton, 2007).

1.3.2 Elevation of utrophin

1.3.2.1 Utrophin – member of the dystrophin/ dystrobrevin family

The UTRN is a homologue of dystrophin, and was first discovered by Love et al. (Love *et al.*, 1989). It is similar to dystrophin as shown in figure 3, and consists of an N-terminal actin-binding domain, a large rod followed by a cysteine-rich domain and the C-terminal region. The gene undergoes promoter-specific expression resulting in 2 known protein variants: UTRNA and utrophin B (Miura and Jasmin, 2006). Both of them have a different unique 5´-exon that splices at exon 3 into a common RNA. In humans, UTRNA is mainly expressed in muscle cells whereas utrophin B is detected in muscle blood vessels. In DMD patients as well as in the mdx mouse, UTRN is found to be up-regulated.

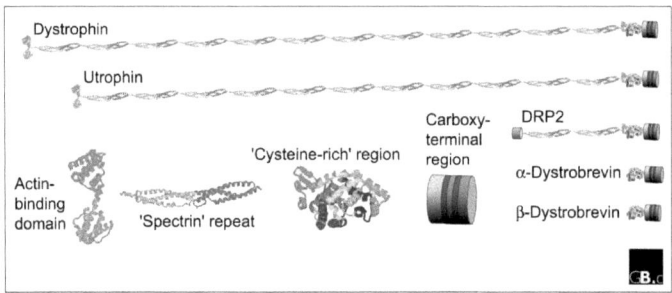

figure 3: Similarities of dystrophin to dystrobrevin and UTRN. Structures of the dystrophin/dystrobrevin family are compiled from the crystal structures of the dystrophin. They also include α-actin-binding domain, 2 spectrin repeats from actinin, and the cysteine-rich region of dystrophin. The structure of UTRN is very similar to dystrophin.
Actin-binding domain: cyan, CH1; green, CH2. 'Cysteine-rich' region: green, WW domain; red, orange, cyan, and purple, EF hands. Carboxy-terminal region: yellow, syntrophin-binding segment; red, leucine heptads. (Roberts, 2001)

UTRN is able to link the extra-cellular matrix to intra-cellular structural proteins in similar a manner as dystrophin and associates with the DGC (figure 4). Both proteins bind indeed the same complement of proteins but differ in their localisation, binding mode and affinity. Whereas dystrophin is found throughout the length of muscle fibres, UTRN is primarily located at the postsynaptic membrane of the neuromuscular junction. However, in regenerating muscle fibres, UTRN is localised throughout the entire sarcolemma as described for early foetal development. Biochemical data demonstrate that UTRN differs in its mode of contact with actin filaments and ß-dystroglycans due to a lack of cluster of basic, actin-binding spectrin repeats. These repeats, which are present in dystrophin, lead to a rearranged actin-binding region in UTRN. Thus, the actin-binding domains are located on different positions in UTRN and dystrophin from which it can be assumed that both proteins bind actin filaments with different affinities. However, a study using stabilised UTRN and dystrophin was able to prove similar affinities for actin filaments. In contrast, binding to ß-dystroglycan was reported to be less effective by UTRN compared to dystrophin. Double knockout mdx mice lacking UTRN demonstrate a more severe phenotype similar to DMD patients and were rescued via adeno-viral delivery of UTRN. In conclusion, elevation of UTRN has become very interesting in connection with the search for a way to treat DMD (Miura and Jasmin, 2006; Ervasti, 2007).

1.3.2.2 Strategies for enhancing utrophin expression

Up to now, an enhancement of UTRN expression can be achieved by 3 different strategies: substitution of the UTRN gene, up-regulation of gene expression using an artificial 3 zinc-finger based transcription factor called Vp16-Jazz (Mattei et al., 2007; Desantis et al., 2009), and delivering or up-regulation of proteins that positively influence UTRN expression. Such a favourable impact was demonstrated for heregulin and neuronal NO synthase (nNOS), respectively. Heregulin, a nerve-derived growth factor, is able to increase UTRN transcription by activation of $GABP_{\alpha/\beta}$ transcription factor complex of its promoter (figure 4). Injections of a small heregulin peptide *in vivo* induced UTRN up-regulation and a significant functional improvement of the dystrophic phenotype in mdx mice (Basu et al., 2007). The myogenic Akt signalling can also promote UTRN expression (Peter et al., 2009). Another mediator for UTRN expression is the neuronal NO synthase that links to the DGC and catalyses the production of nitric oxide (figure 4). The expression of nNOS is reduced at the sarcolemma in dystrophin-deficient muscle, and it is thought to function as a protector against muscle damage. A study in mdx mice overexpressing nNOS transgene in the dystrophic muscle provided compelling evidence that nNOS up-regulation can diminish the dystrophic pathology. Muscles from transgenic mdx/nNOS mice showed improvements in muscle membrane injuries associated with a large decrease in serum creatine kinase concentrations. Substitution of L-arginin, the substrate of nNOS, is reported to increase nNOS expression in mdx mice as well. But the mechanism of the effect of L-arginin has not been discovered yet. The up-regulation of UTRN is discussed to occur at transcriptional level as a result of inhibited calpain-mediated UTRN degradation (figure 4) (Miura and Jasmin, 2006). But UTRN up-regulation can also be stimulated via signalling pathways that involve calcineurin and nuclear factor of activated T-cells (NFATc) as key components (figure 4). One of the main targets of calcineurin is the NFATc. Calcineurin de-phosphorylates NFATc, and this leads to its activation and migration into the nucleus where it can interact as transcription factor in concert with other factors such the nuclear respiratory factor 2, *ets*-related GABP$_\alpha$/β transcription factor complex, and PGC1-α on the UTRN promoter (figure 4) (Angus *et al.*, 2005). These data imply that UTRN expression is activated by signalling pathways with calcineurin and NFATc as key components. *In vivo* experiments using calcineurin*/mdx mice over-expressing calcineurin represent an induction of UTRN expression with a concurrent, marked alleviation of the dystrophic pathology (Chakkalakal *et al.*, 2004). Interestingly, over-expression of UTRN at high levels does not lead to toxic consequences throughout a range of tissues in mdx mice (Fisher *et al.*, 2001).

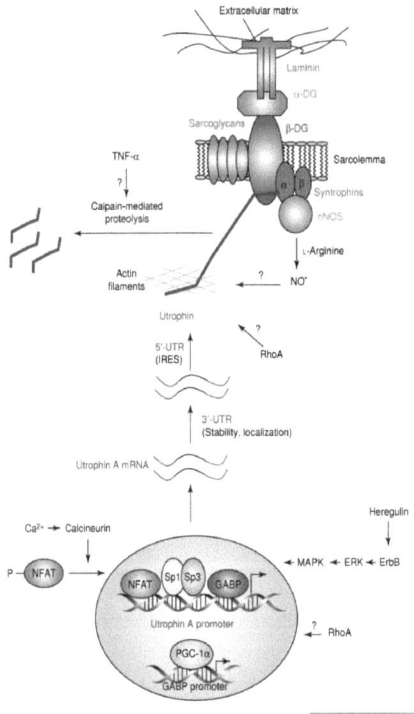

figure 4: Expression and regulation of UTRN. Expression of UTRN, in particular UTRN A, is mainly restricted to the postsynaptic sarcoplasm in fast and slow muscle fibres of normal skeletal muscle. Similar to dystrophin, UTRN provides a link between the actin cytoskeleton and the extracellular matrix via the DGC at the sarcolemmal resulting in a downstream signalling cascade as shown for dystrophin (simplified figure, more details are shown in figure 5). Expression of UTRN is mediated by several transcription factors. The transcription factor complex of GABPα and GABPβ is promoted by a cascade that is triggered by ErbB receptors bound to heregulin. The transcription of GABPα can in turn be initiated by PGC-1α. Many other transcription factor recognition sequences are found at the UTRNA promoter including for NFATc, Sp1 and Sp3. The last 2 are ubiquitously active in numerous genes. NFATc is activated by de-phosphorylation and can also mediate UTRNA transcription. RhoA has been found to be involved in UTRN transcription in an unknown manner, but confers stability to UTRNA. Stability and localisation of the UTRN transcript is given through the 3' untranslated region. The 5'UTR stimulates UTRNA translation for muscle regeneration. Calpain-mediated degradation of UTRN is thought to be induced by TNFα. The protein nNOS and its substrate are found to stimulate UTRN exression via nitric oxide (NO). (Miura and Jasmin, 2006)

Since UTRN expression can be stimulated by activation of the calcineurin/NFAT pathway (figure 4), an endogenous up-regulation using pharmacological compounds could also be considered, but is often less efficient than a gene transfer-mediated UTRN up-regulation. The main difficulty with this strategy is to achieve a restriction of UTRN expression at the neuromuscular junction and a widespread expression in myofibers. Among the pharmacological approaches glucocorticoids such as deflazacort are especially effective to slow down the progression of DMD (Angelini, 2007; Houde et al., 2008; Manzur et al., 2008a). Although a therapeutic effect is visible, the molecular basis of the pharmacological effect is not well understood and is mainly assumed to be anti-inflammatory. Furthermore, the potent immunosuppressive substance cyclosporin A (CsA) was reported to benefit muscular dystrophy once (Sharma et al., 1993). However, contrary data resulting from the use of mdx mice were also reported (Stupka et al., 2004). In addition to anti-inflammatory effects, it is well known that the calcineurin inhibitor CsA is involved and that deflazacort is suspected of being involved in the NFATc signal transduction pathway (figure 4) (St-Pierre et al., 2004; Stupka et al., 2004). Investigations using mdx mice indicated an activation of the calcineurin phosphatase associated with an up-regulation of the NFATc target gene UTRN using deflazacort (St-Pierre et al., 2004). However, an increased activity of the NFATc/calcineurin pathway has to be restricted to the muscle suffering from dystrophic pathology since cardiac hypertrophy is associated with induced calcineurin (Molkentin et al., 1998).

1.3.3 Dystrophin downstream strategies

1.3.3.1 Dystrophin glycoprotein complex

Dystrophin functions as a connector between the cytoskeleton and the sarcolemma of muscle cells. But the expression of dystrophin is not limited to the skeletal and cardiac muscle. Truncated isoforms are also found in kidneys, neurons, lymphoblastoid, brain, muscle, Purkinje cells, the retina, and in Schwann cells. Interestingly, the shortest forms Dp70 and Dp40 are present ubiquitously (Ervasti, 2007).

Dystrophin is part of the DGC faced at the cytoplasmic membrane of muscle fibres. The DGC contains both transmembrane components such dystroglycan and sarcoglycan as well as cytoskeletal proteins such as dystrophin and syntrophin (figure 5) (Ervasti and Sonnemann, 2008). The ß-dystroglycan is the key component of the DGC. It is bound to the cystein-rich domain of the dystrophin and to the sarcoglycans, and provides a direct link to laminin-α2, an extracellular matrix protein. The sarcoglycan sub-complex consists of α-, β- ,δ-, ε-, γ- and ζ- sarcoglycans. A loss of one of these sarcoglycans has been found to be associated with autosomally inherited types of muscular dystrophies, such as the Limb girdle muscular dystrophy form 2C, and causes a similar phenotype to DMD. The sarcoglycans are found to function for plasma membrane permeability. The tetraspanin family member sarcospan is associated to the sarcoglycan facing the cytoplasm. But its function remains elusive. Another sub-complex within the DGC is formed by the 5 syntrophins. The syntrophins are connected to the C-terminal domain of dystrophin. Additionally, syntrophins can also interact with phosphatidylinositol bisphosphate (PIP2), nNOS, calmodulin and the growth factor receptor-bound protein 2 (Grb2). The latter activates a complex downstream signalling. Some domains may also able to influence sodium channels.

Other components of the DGC are the dystrobrevins, which own a cysteine-rich and a C-terminal domain homologous to those found in dystrophin (figure 5). The 2 isoforms α- and β-dystrobrevin interact with dystrophin and UTRN through the coiled-coil domain. Because of the similarity to dystrophin, dystrobrevins are thought to be a target for therapeutic strategies. Other proteins connected to the DGC are dysferlin, calpain-3, caveolin-3, and filamin-C (Ervasti and Sonnemann, 2008).

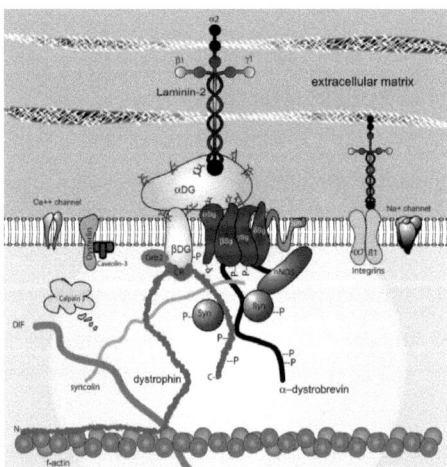

figure 5: Components of the dystrophin-glycoprotein-complex (DGC). Dystrophin mediates conjunction of the contractile elements in the cell with the extracellular matrix via the dystroglycan complex. The DGC consists of α-, β- ,δ-, ε-, γ- and ζ- sarcoglycans (SG; sarcogylan complex), sarcospan (Spn), 5 syntrophins (Syn), α- and β-dystrobrevin (aDG, bDG). Laminin-2 is an extracellular ligand, whereas F-actin and desmin-intermediate filaments (DIF) are intracellular binding partners. The proteins dysferlin, calpain-3, caveolin-3, syncoilin (S), and filamin-C are connected to the DGC, and connect the complex to signalling proteins as Grb2 and nNOs that initiate a dystrophin downstream signalling cascade. (Odom et al., 2007)

1.3.3.2 Molecular consequences of the lack of dystrophin

Usually, the absence of dystrophin is associated with DMD while truncated forms of dystrophin are correlated with the less severe phenotype of BMD. But, exceptions have been reported in both directions suggesting the existence of pathways compensating the dystrophin defect (Winnard *et al.*, 1993; Prior *et al.*, 1997; Hattori *et al.*, 1999; Pradhan, 2004; Ferreiro *et al.*, 2009).

The lack of dystrophin leads to a failure of incorporation of the DGC to the sarcolemma. Consequently, dystrophin downstream cascades signalling trails away or cannot be established. A dramatic increase in intracellular calcium is observed, suggesting that this may be a result of physical sarcolemmal breaks or calcium channel leak. In dystrophic muscle cells, it was demonstrated that calcium-permeable channels can exhibit an increase of activity resulting in calcium-activated proteases as well as modulation of other intracellular regulatory proteins and signalling pathways. Calcium channel blockers, e.g. diltiazem and verapimil, are used to reduce the calcium content in the diaphragm and cardiac muscle in mdx mice (Matsumura *et al.*, 2009). But a significant positive effect of these drugs in DMD patients cannot be evidenced (Phillips and Quinlivan, 2008). In this context, feasible functions of dystrophin and the DGC are discussed including regulation of calcium ions, storage of calcium channels or intracellular calcium, or an indirect alteration of calcium regulation caused by membrane tearing (Hopf *et al.*, 2007). Another characteristic of the disease is the limited capacity of DMD myoblasts to grow that is directly related to the progressive muscle degeneration (Blau *et al.*, 1983b). Concomitantly, muscle cells in culture are found to be fully capable of initiating myogenesis and differentiation (Blau *et al.*, 1983a). As investigated in this group, the cyclin-dependent kinase inhibitor 1A (p21), which negatively regulates G1 to S progression, is increased in DMD patients, and is thought to be related to decreased myoblast proliferation (Endesfelder *et al.*, 2000). Transfection of ASO and short interfering RNA (siRNA), respectively, against p21 exhibited an increase of proliferation in dystrophin-deficient myoblasts, and provides a potential therapeutic strategy, which makes allowances for proteins downstream dystrophin (Endesfelder *et al.*, 2003; Endesfelder *et al.*, 2005). For grasping an extended range of genes that are involved in DMD pathology, differences of total transcriptomes or proteomes were investigated (Bakay *et al.*, 2002; Haslett *et al.*, 2002; Noguchi *et al.*, 2003; Haslett *et al.*, 2003; Baker *et al.*, 2006; Dogra *et al.*, 2008). This group has reported about differences between transcriptomes in the very rare case of 2 brothers with an intra-familially different course of DMD with the younger brother being far less affected than the elder one when compared at the same age. To isolate candidate genes responsible for the milder course, a cDNA-library enriched with transcripts over-expressed in the younger brother was constructed. A PCR-based subtraction hybridisation technique was used for the library. Thereby, 11 transcripts were identified as differentially expressed at which elevated expression levels in the DMD patient with mild pathology has been confirmed by qRT-PCR for casein kinase 1 alpha 1 (CSNK1A1), ras related protein 2B (RAP2B), dynactin 3 light chain (DCTN3), myosin light chain 2 (MYL2) and core binding factor beta (CBFB), and an hypothetical gene compared to the elder brother (Sifringer *et al.*, 2004). Despite this small number of differences, the problem reveals that a correlation of discovered genes to dystrophin as well as to each other can apparently not be discovered. In experimental studies concerning transcriptome and proteome analyses of DMD patients compared to normal individuals, this aspect has not been considered. However, the reports usually result in a list of discovered genes and proteins rather than conducting the advancing step of their integration in dystrophin associated signal transduction pathways.

To meet interacting requirements, a network has to be established including primary knowledge of dystrophin downstream cascades and experimentally discovered genes. The new arising field of computational systems biology provides appropriate approaches for modelling, analysing as well as animation of such signal transduction networks. Thus, it offers an expert insight into molecular consequences of the lack of dystrophin for the development of new experimental applications and consequently dystrophin downstream therapy strategies.

The present thesis applies theoretical modelling techniques to

(1) summarise pathway information of dystrophin downstream processes into one model including visualisation and animation facilities

(2) identify important parts of the network through theoretical knockout analyses

(3) design new experiments based on theoretical results.

Computational systems biology

1.4 Interest of systems biology

The field of biology, molecular biology in particular, has changed considerably over the last few years. Traditionally, molecular biology was differentiated in genomics, gene expression and protein analysis and thus investigated individually one after another. But, biological systems are very complex networks and function rather as a unit by interaction of gene expression, metabolic pathways as well as cell internal and external communications than autonomously. As a selected and restricted example, figure 6 demonstrates the multifaceted interplay of filaments and macromolecular complexes in cell interior of the Dictyostelium discoideum cell (Ellis and Minton, 2003). The figure represents a 3-dimensional description without the need of functional interaction of depicted macromolecules.

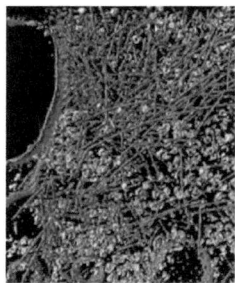

figure 6: Cryoelectron tomography of crowded cell interior of the Dictyostelium discoideum cell. The 3-dimensional picture is a reconstruction of an image series taken from a range of tilt angles. The actin filaments are shown in orange; grey complexes represent the ribosome and other macromolecular complexes. The membranes are illustrated in blue. (Ellis and Minton, 2003)

Functional aspects of components of biochemical networks are illustrated in figure 7 serving as a model of the dynamic processes in the cell including DNA, RNA and protein level as well as intercellular signals, and of which influence on each other (Arbeitsausschuß Bioinformatik der DECHEMA e.V., 2006). In addition, the development of new techniques such as micro-array analysis and sequencing facilitate the production of high-throughput and large-scale multi-dimensional data sets.

Altogether, it shows the importance of gradually extending and integrating all these information in a hierarchically structured model, which represents all the data in its biological system rather than considering them separately and in a reduced state. This inspired the development of new computational and mathematical approaches for analysing biological information and for network modelling. Accordingly, biology has changed into a cross-disciplinary field including biochemists, computer scientists, engineers, mathematicians, bioinformaticians, and physicists to understand biological organisms in their entirety.

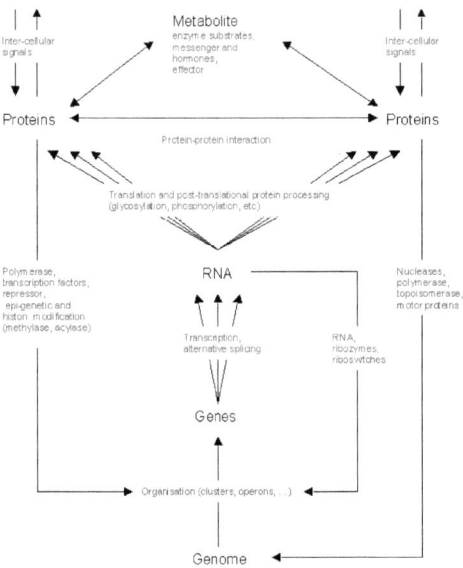

figure 7: Gene expression is regulated at different hierarchical levels and influenced by numerous positive and negative feedback mechanisms resulting in amplification of signal- and product-variety. Hundreds of thousands of proteins are encoded by approximately 30,000 genes of the human genome. These proteins correspond to a complex interaction network, which finally enables the function of an organism. Examples and more detailed explanations are in red.
(after (Arbeitsausschuß Bioinformatik der DECHEMA e.V., 2006))

The study field combining biological high-throughput data sets, specific methods of analysis and mathematical modelling is called systems biology and involves extensive experiments and hypothesis-driven modelling. Challenges of computer science and mathematics in systems biology take account of full information content of the genome, extraction of protein and gene regulatory networks from the genome. Integration and visualisation of high-throughput and multi-dimensional data sets as well as modelling of dynamic mathematics converted from static networks ensue. Additionally, prediction of protein structure and function, identification of different cellular conditions by associated key components, building of hierarchical models as well as the reduction of the complexity of multi-dimensional models are further benefits of systems biology (Baxevanis and Ouellette, 2005). Consequently, the most important methods in system biology are databases, mathematical models for quantitative as well as qualitative networks, and derived computer simulations. Data can be used to express a hypothesis, and following models are applied to design new experiments to verify the hypothesis. Another important aim of theoretical modelling is the reduction of time and reagent consuming experimental approaches, animal studies, and is in particular of interest for complex events. Cell cycle and apoptosis are first examples. In case of the cell cycle, reports represent mathematical models concerning G1/S transition and G1-phase (Swat *et al.*, 2004). For apoptosis, a mathematical model was reported that identified an implicit positive feedback of 2 apoptosis mediating proteins if the same proteins are suppressed by inhibitors of apoptosis (Legewie *et al.*, 2006). A so far unreachable aim of systems biology is the model of a virtual cell/organism enabling the performance of experiments *in silico* such as the prediction of the effect of medication as well as phenotype/genotype correlation in diseases (Baxevanis and Ouellette, 2005).

1.4.1 Theoretical methods in systems biology

In systems biology, a distinction is made between quantitative and qualitative modelling. Quantitative networks are mainly used for metabolic pathways. These networks take advantage of kinetic data, and are usually based on Michaelis Menten kinetics, its relevant stoichiometry, substance fluxes as well as the steady state assumption. In contrast, because of the lack of quantitative data, qualitative networks are primarily applied to model signal transduction pathways and gene regulation. These pathways do not rely on kinetic data, substance fluxes and the steady state assumption, but present fluxes of information or signals. Additionally, in most qualitative approaches time properties are not considered.

1.4.1.1 Quantitative modelling

One of the main applications of systems biology are quantitative metabolic networks since for these pathways kinetic data are available. On the basis of extensive kinetic data, these networks interpret static metabolic systems in dynamic models, and can be expressed mathematically. An established mathematical approach to describe the control in metabolic pathways is called metabolic control analysis (MCA) (Kacser, 1995; Heinrich and Schuster, 1998). MCA is an ODE-based (Ordinary Differential Equation) computational method to study changes in fluxes and concentrations of metabolic networks. It considers the effects of the different enzymes involved by which the managing reaction steps can be examined (Salter et al., 1994). Hence, MCA provides information for changes in the metabolic pathway to increase or decrease reaction processes, for example in the case of biotechnological production procedure. System behaviour can be described by the control coefficient and the elasticity coefficient. The control coefficient reflects the system property of an enzyme expressing the differential alteration of the substance flux (resulting in a steady state) by metabolic pathways through differential modification of the activity of involved enzymes. The elasticity coefficient is the local property of an isolated enzyme in terms of changes in metabolite concentrations. The control coefficient and the elasticity as a result of MCA (a form of sensitivity analysis) require data of the model structure and its biochemical parameters. This control can be determined by application of a perturbation of the enzyme of interest and evaluation of system variables after the system is returned to steady state. The enzyme kinetic parameters are usually examined *in vitro* including substance flux and intracellular metabolite concentrations (metabolic profiling) (Link and Weuster-Botz, 2007). MCA, which is usually used for metabolic pathways, can also be applied to qualitative modelling to describe signalling and genetic networks. Based on experimental data, theoretical effects of divergence and convergence of selected genes in signalling cascades are of particular interest in this case. They often display, for example, 2fold series of connected inhibition in feed-forward paths, and multiple activations or blocking of feedback-loops.

1.4.1.2 Qualitative modelling

Qualitative modelling works at another abstraction level than modelling of quantitative networks. Graph theoretical analysis and network animations are also essential at this point. Stoichoimetry, time properties, and kinetic parameters of these processes to be modelled are often unknown. For successful parameter estimation, a critical amount of kinetic data must be available. Consequently, a classic kinetic modelling using ODEs is non-applicable. Hence, to get nevertheless insight into the system behaviour a qualitative rather than a quantitative model can be applied, but would be similar to a static form (Wildermuth, 2000; Cascante et al., 2002; Visser and Heijnen, 2002; Arbeitsausschuß Bioinformatik der DECHEMA e.V., 2006).

An important purpose of qualitative modelling are medical applications that include gene regulatory parameters and complex protein interaction rather than kinetic data and are becoming more and more relevant in order to reduce experiments for target finding (Hood *et al.*, 2004). For gene regulatory networks that consider such gene regulation as switching of and switching on, a Boolean network would apply. Transcription and signalling processes in addition to gene regulation require a mathematical formalism to include different abstraction levels in one model. Various abstraction levels are enabled by classic Petri nets that are discrete models, and do not rely on kinetic data. There exist different Petri net classes for modelling systems of different description levels, e.g. hybrid Petri nets. For biochemical systems, hybrid Petri nets are used when some kinetic data are known. Hybrid Petri nets include qualitative as well as quantitative characteristics, and has been used for modelling of metabolic networks, signal transduction networks, and gene regulation networks (Matsuno *et al.*, 2000; Matsuno *et al.*, 2003; Saito *et al.*, 2006).

1.4.2 Petri net basics

In 1962, Petri nets were defined by Carl Adam Petri for modelling of systems with concurrent processes. Petri nets are discrete models of information and object fluxes to describe systems and processes at different abstraction levels. Thus, it does not rely on kinetic data. A Petri net is a directed, bipartite and labelled graph. In addition to its graphical representation, its animation facilities afford a simple and intuitive evaluation of the biological behaviour of the model. Hierarchical elements serve structuring of the complex system (Baumgarten, 1996). Formal definitions for Petri nets and invariant analyses and their applications approaches are summarised in the glossary (section 9.1).

The Advantages of Petri nets are

a unique modelling language for different abstraction levels,

modelling of sequential and mutually excluding and causally independent activities,

formal analysis, verification and simulation techniques supported by computer software tool,

graphical depiction using comprehensible modules (Baumgarten, 1996).

Composition of a Petri net

Petri nets consist of places (P) and transitions (T), which are connected by directed arcs. Because of the bipartite property, a direct link between transitions and transitions, and places

INTRODUCTION

and places is not allowed. Places are drawn as circles and represent passive system elements e.g. conditions, states or in biological networks proteins, genes, transcriptions factors etc. Transitions as rectangular figures (or squares) symbolise the active part of the system such as events or biochemical reactions. Transitions without pre-places (post-places) are called input (output) transitions (here drawn as flat rectangles). Connecting arcs correspond to causal correlations of passive and active system elements. A transition is connected to pre-places, which represent pre-conditions via its incoming arcs. Accordingly, post-places that form post-conditions are connected via its outcoming arcs. Places may contain tokens, depicted as dots within places. Tokens are movable objects, and can represent a number of molecules or chemical compounds. The distribution of any number of tokens is called marking (m) and represents a certain system state. The initial marking defines the primary state of the system. The arc weights specify the number of tokens being produced or used up if an event takes place. The event of a transition is triggered if all its pre-conditions are fulfilled. Thus, the transition is enabled, and has concession to fire.

The firing rule describes the dynamic behaviour of the system. A classic discrete place/transition net (P/T net) exhibiting a timeless firing rule is considered. A transition has concession to fire if all pre-places contain at least as many tokens as the transition charges for the event as indicated by the corresponding arc weights (Baumgarten, 1996). Examples for the firing rule of a P/T net are exemplarily demonstrated in figure 8a-c. Only the event of the transition in figure 8a is triggered because the required number of tokens demanded by the arc weights is available in the pre-place in contrast to figure 8b. After an event, all post-places contain at least as many tokens as forced by arc weights after the transition as shown in figure 8c.

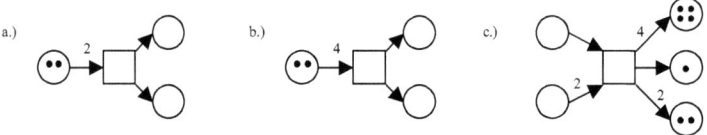

figure 8: Demonstration of the firing rule for P/T nets: a) The pre-condition of the transition is fulfilled as designated by the number of tokens included in the pre-place (2) and the arc corresponding weight (2). b) The transition has no concession to fire since 4 tokens are required as represented by the corresponding arc weight (4). c) The post-places receive the number of tokens as specified by arc weights, after the transition fired.

Read arcs

Especially in signal transduction pathways, enzymes or cytoskeletal proteins participate in a products-releasing reaction, but are not used up and consequently may return to the system for the next reaction. This special kind of reaction is realised using read arcs, drawn as a bi-directed arc. As a result, the number of tokens on the pre- and post-places does not change through firing of the considered transition.

Analysis approaches

Since Petri nets play a major role in many technically as well as biologically oriented fields (e.g., computer science, analysis of distributed systems, production processes, and molecular biology), various analysis methods have been developed in the last few decades to analyse the net behaviour of these systems with concurrent processes.

For the analysis of biological networks, particular network properties are important and have to be defined, e.g. liveness, reachability, support of a vector, invariants, etc. The meaningful liveness of a Petri net is described at different levels and indicates a continuously working network. The liveness of a transition exists if it is enabled in any reachable state, and can be enabled if it is enabled in at least one reachable marking. The marking of a Petri net is called reachable if a firing sequence of transitions exists which turns the initial marking back into this marking. Finally, the Petri net whose initial marking may be reachable by any achieved marking again, is defined as reversible (Baumgarten, 1996).

For Petri net approaches in biology, a number of analysis techniques are considered that are applied to this thesis: invariant analysis, maximal common transition set analysis, cluster analysis, and Mauritius map analysis.

1.4.3 Invariant analysis

Invariant properties of the system are valid for each initial marking and each reachable state. Invariant analysis is a primary study to explore firing properties as a result of the structure. The computation of invariants is based on the incidence matrix that corresponds to the stoichiometric matrix in metabolic networks. It is distinguished between place and transition invariants. In further studies, transition invariants (t-invariant) are explored (Starke, 1990). T-invariants represent multi-sets of transitions, their firing, and frequency of firing. Within a t-invariant, firing of all included transitions with the firing frequency as indicated in the *Parikh vector* leads to the same marking as before firing. The invariant property can be calculated as nontrivial, semi-positive integer solutions of linear equations systems

(Lautenbach, 1973), and is independent of any firing in system behaviour. In biology, this applies to the steady state assumption for metabolic systems.

The description of an incidence matrix is exemplarily demonstrated in figure 9. The solution vector defines a *Parikh vector*, and is written as a column vector. It indicates firing frequencies for each transition in the t-invariant to reproduce a given marking.

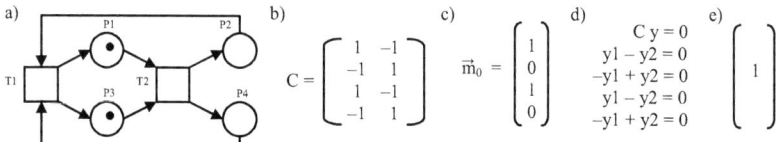

figure 9: The state of a Petri net, the defined incidence matrix, and the vector describing the initial marking.
a) The Petri net is live. b) The incidence matrix describes the topology of the Petri net in a.), but not the initial marking. The 2 columns indicate transitions T1 and T2 and the 4 places P1...P4 are specified by the rows. For example, position p1 and T1 shows 1 because one token is given to P1 by firing of T1. In contrast, position P1 and T2 presents −1 since a token is taken by firing of T2 from P1. c) The initial marking m_0 presented in vector form. It indicates the number of tokens on each place. d) The linear equation system of the Petri net for the determination of p-invariants (rows). A second linear equation system is setted up for the determination of t-invariants (column). e) The solution for each linear equation is 1.
[modified after http://osg.informatik.tu-chemnitz.de/lehre/old/ss07/vs/12-petrinets.pdf]

As shown in figure 9, a Petri net can be represented as incidence matrix and a vector for initial marking. Linear equation systems have an infinite solution space. Therefore, minimal solutions are of interest. Each possible solution can be computed by linear combination of minimal solutions or by multiplication with a positive integer number. Minimal invariants are considered in this thesis. Criteria for validation of a biochemical network have been formulated.

Minimal validation criteria for a biological Petri net include a connected net that is covered by biological meaningful t-invariants. In case of every transition is integrated in at least one t-invariant, the Petri net is covered by t-invariants (CTI). An uncovered Petri net contains at least one transition that is not integrated in a t-invariant and may be excluded without affecting the system behaviour. A t-invariant defines a connected subnet, consisting of its support, its pre- and post-places, and all arcs in between. Thus, it describes the minimal basic system behaviour. In molecular pathways, each invariant of a validated Petri net model needs to fulfil a biologically correct and meaningful function (Heiner and Koch, 2004; Koch and Heiner, 2008). Another tool for identification of subnets within biological network are elementary flux modes, which correspond to t-invariants (Schuster *et al.*, 1999).

1.4.4 Maximal common transition set

The t-invariant analysis provides the opportunity to decompose a network into basic pathways to obtain clarity, which is helpful for studying the network. But the number of t-invariants increases exponentially with the number of transitions and arcs in a network model. In complex and large Petri nets, the manual study of all t-invariants is very intricate. Consequently, a possible way of handling such large numbers of t-invariants is to decompose the network into such sub-nets as for example in the case of the transition are subsumed of transitions according to their exclusive co-occurrence in t-invariants The summarised number of t-invariants is called maximal common transition set (MCT-set). For example, transitions of an MCT-set are located in the same minimal t-invariants in any case. This relation results in maximal sets of transitions. The MCT-sets can define disjunctive subnets, which may represent functional units or, in biological networks, a defined biological function. The transitions of an MCT-set always occur together. Thus their up- and down-regulation should be relatively.
Altogether, MCT-sets decompose nets into disjunctive subnets, which are not necessarily connected, whereas t-invariants describe subnets, which are always connected (Sackmann *et al.*, 2006).

1.4.5 Cluster analysis

Cluster analysis techniques are well-established approaches to explore enormous amounts of data. Thus, this is another way of decomposing extensive biological pathway models is represented by the clustering technique. It is used to cluster t-invariants in biologically functional sub-units into t-cluster, which, unlike MCT-sets, can overlap. A cluster analysis can be seen as a 3 step process, encompassing the following main steps: (1) selection of a distance measure to compute the distance between all pairs of objects, (2) selection of a clustering algorithm to group the objects based on the computed distances, and (3) selection of a cluster validity measure to identify the optimal number of clusters for interpretation. A detailed description of each of these steps is given below (Grafahrend-Belau *et al.*, 2008a). The distance measure used in this work is based on the work is the *Tanimoto* coefficient based on the *Parikh vector*. The method "Unweighted Pair Group Method with Arithmetic mean"

(UPGMA) is a technique for hierarchical clustering of tree construction and was initially developed for calculating phylogenies (Backhaus *et al.*, 2000).
For more detailed information see Grafahrend-Belau et al. and Sackmann et al. (Sackmann *et al.*, 2007; Grafahrend-Belau *et al.*, 2008a; Grafahrend-Belau *et al.*, 2008b).

1.4.6 Mauritius map

A graphical representation of sets belonging to the support of t-invariants is called Mauritius map. An application of Mauritius maps is modelling the Petri net as a tree structure. Primarily, alterations in network behaviour are analysed when transitions are knocked out. This is of particular interest in biological models.

Mauritius maps are defined as finite binary trees as widely used in computer systems. The root of the tree is located in the lower left corner and only branches out to a right sub-tree, but not to a left sub-tree. For Mauritius maps, the Petri net model needs to be covered by t-invariants since all uncovered transitions are excluded from the right sub-tree (Heiner and Koch, 2004; Koch and Heiner, 2008). A transition of a t-invariant corresponds to a vertex. Vertices belonging to the same t-invariant are connected by horizontal edges but also by vertical edges from the left sub-tree to the right upper sub-tree. The next t-invariant is signalised by a branch in the tree. Sub-networks are represented in interior vertices providing left and right sub-trees and consist of all intermediate vertices from the root to the considered vertex. Parent vertices are indicated by a junction of horizontal and vertical lines, which depict the edges of children vertices.

The distance of a transition and the respective vertex to the root gives an idea of the relevance of that transition for the Petri net behaviour. The closer the vertex is located to the root the higher the impact of the considered transition for the network model. Transitions located on the horizontal line that starts from the root are characterised as the most important transition having the highest impact in case of a knockout. This applies to transitions occurring in all t-invariants. The number of t-invariants lost by initiation of a knockout of a single transition specifies the impact of that transition. All pathways described by the corresponding right child and subsequent vertices are affected, i.e. destroyed, whereas the left child and all parents are not influenced. Consequently, the sub-net (left child) without the knocked out transition remains active (biological functional). In contrast, the biological function of the other sub-net (right child) containing the knocked out transition is discontinued, and according to this, its function completely depends on that transition.

The following figure 10 illustrates a Mauritius map (figure 10b) of a simple signal transduction pathway (figure 10a), which can be visually explored. The signal enters t_{in} and may reach t_{out} by passing t_1 and t_2 or alternatively, transitions t_3 and t_4. Thus, 2 t-invariants describe the behaviour of this small net: (1) with (t_{in}, t_1, t_2, t_3 and t_{out}) and (2) with (t_{in}, t_3, t_4 and t_{out}). Obviously, transitions t_{in}, t_3, and t_{out} exclusively occur in both t-invariants, and represent an MCT-set. Consequently, a lack of one of them would result in a dead Petri net. For a knockout analysis, a transition needs to be chosen that is only a member of one of the t-invariants such as t_1, t_2 and t_4. For example, the knockout of transition t_1 leads to a loss of approximately 50% of the net (horizontal line). The functional part is represented by the left child at the end of the vertical line, which can be deleted by knocking out t_4.

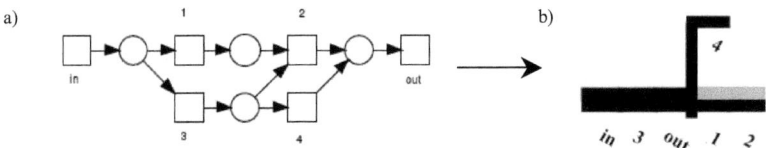

figure 10: A simple Petri net (a) and the corresponding Mauritius map (b).
After a signal is initiated by t_{in}, it can either pass transitions t_1, t_2 or t_3, t_2 or t_3, t_4 and is released by t_{out}. Thus, the signal flow is described by two t-invariants: (1) with (t_{in}, t_1, t_2, t_3 and t_{out}) and (2) with (t_{in}, t_3, t_4 and t_{out}). Transitions t_{in}, t_3 and t_{out} are present in both t-invariants and result in an MCT-set. Their knockout would lead to a dead Petri net. By choosing transition t_1, one of the t-invariants would lose its function whereas the left child at the end of the vertical line remains. The second t-invariant can be knocked out by deletion of t_4. (modified after (Grunwald et al., 2008))

1.4.7 Application of Petri nets to biochemical systems

In systems biology, an ever increasing number of metabolic (Reddy et al., 1993; Koch et al., 2005; Chen and Hofestadt, 2006) and signal transduction pathways (Heiner et al., 2004) have been modelled as Petri net networks. For example, in biochemical networks, the sucrose breakdown in the potato was modelled using Petri nets (Koch et al., 2005). The authors demonstrate a qualitative analysis for structural validation of metabolic networks. A hybrid Petri net has been applied to the operon model and the growth pathway control of λ-phage (Matsuno et al., 2000). The use of Petri nets can be broadened for simulation of time-resolved applications as shown for the time-dependent somatic complementation in a model describing the commitment to sporulation regulated by far-red light and glycose in Physarum polycephalum (Marwan *et al.*, 2005). A technique for modelling qualitative gene regulatory networks using Petri nets is reported by Chaouiya (Chaouiya *et al.*, 2004). A comprehensive overview of biological networks modelled as Petri nets is given by Chaouiya (Chaouiya, 2007).

The dynamics of metabolic systems can be investigated by simulation at which appropriate kinetic data required for such simulations are established. But for signal transduction pathways, kinetic parameters often are not available. Thus discrete models are used. These discrete models are considered with or without timing. Another high level of Petri nets are coloured Petri Nets (CPN), which apply to complex models to describe them in a manageable way. In CPN, the tokens on a place can be distinguished. They represent a refinement of p/t nets. CPN also provide formal analysis. The usefulness of CPN was demonstrated in studying and analysing the dynamic behaviour of the epidermal growth signal transduction pathway (Lee *et al.*, 2006).

The applications of metabolic and signal transduction networks demonstrate transferability of well-known pathways to Petri net models. Up to now, applications of Petri nets to signal transduction pathways involved in pathogenesis of human neuromuscular disease, in particular DMD, have not been reported. For molecular changes in DMD, neither quantitative nor qualitative models intended to include protein interactions, transcription factors, and gene expression regulations have been developed so far. But this is important as pointed in section 1.3.3.2.

2 Conceptual formulation

The thesis is based on a study of 2 siblings with an intra-familially different course of DMD. Using a cDNA subtraction library, 5 genes were determined which are up-regulated in the DMD patient with the mild phenotype and therefore may be responsible for the less severe clinical course of the disease.

The first objective of this work is to develop a signal transduction network that involves these candidate genes and elucidates an association to dystrophin.
This includes the following sub-items:
Establishing of a signal transduction pathway that connects dystrophin downstream cascades with genes identified using the cDNA subtraction library.
Remodelling of the signal transduction pathway into a Petri net.
Structural analysis of the network behaviour for verification of systems consistency and biological correctness.
Theoretical knockout analysis to study functional dependencies in the model and to determine the most important components of the net for the network behaviour and consistency. These candidate genes/proteins may also have a great impact in cellular and biological studies.

The second objective is to conduct experimental studies to verify the supposition derived from Petri net analyses. That includes investigations of gene expression and modulation of selected proteins of the network.
This involves the following studies:
- Sequencing of a putative promoter region in phenotype-modifying candidate genes using both siblings studied by the cDNA subtraction library.
- Determination of mRNA expression levels of selected genes of the network using 5 classic DMD patients, the mild DMD patient and 5 normal controls.
- Modulation of the activity and/or expression of proteins of the network, which are chosen through the Mauritius map analysis, using SkMCs of DMD patients and normal controls in cell culture.
- Expression and proliferation analysis of modulated SkMCs at mRNA and protein level of selected genes.

3 Material and Methods

Molecular biology

3.1 Material

3.1.1 Chemicals

Designation	Source
Agarose ultra pure	Life Technologies
Ammonium acetate	Merck
Ammonium hydroxide solution	Fluka
Ammonium sulfate (for analysis)	Merck
Ampicillin	Sigma
Antioxidant NuPAGE (Western blot)	Invitrogen
APS	Serva
beta-Mercaptoethanol	Roth
Bromphenol Blue	Serva
Bovine serum albumine	Gibco BRL
Calcium chloride	Merck
Chloroform	Roth
Cyclosporin A	Merck
Deflazacort	Sanofi Aventis
DEPC	Fluka
Dithiothreitol	Invitrogen/ Promega
DMSO	Roth
dNTP-set	Invitrogen
EDTA (Titriplex III)	Merck
Ethanol	Merck
Ethidium bromide	Merck
Formaldehyde	Merck
Formamide	Merck
Glycine	Serva
Glycogen	Roche
Hexanucleotide pd(N)$_6$	Invitrogen/ Promega
IC261	Merck
Isopropanol	Merck

Designation	Source
Skim milk powder	Merck
Magnesium chloride (for PCR)	Applied Biosystems
Magnesium chloride	Merck
Methanol	Merck
Neomycin	Biochrom KG
Okadaic acid	Merck
Paraformaldehyde	Merck
PBS Dulbecco (50x	Biochrom KG
Penicillin	Biochrom KG
Phenol/Chloroform/Isoamylalcohol	Roth
Potassium chloride	Merck
Potassium dihydrogen phosphate	Merck
ProLong Antifade	Invitrogen
Reducing agent NuPAGE (Western blot)	Invitrogen
RNase free water	Ambion
Sodium acetate	Merck
Sodium chloride	Merck
Sodium citrate	Merck
Di-Sodium hydrogen phosphate dihydrate	Merck
Sodium hydroxide	Fluka
Sodium hydroxide solution (50%)	Merck
SDS (Sodiumdodecylsulphate)	Serva
Streptomycin	Biochrom KG
TaqMan Universal PCR Master Mix No AmpErase UNG	Applied Biosytems
Triethanolamine	Merck
Tri Pure	Roche
Tri Reagent	Sigma Aldrich Chemie
Tris-(hydroxymethyl)aminmethan	Merck
Trypan blue	Sigma Aldrich Chemie
Trypsin/ EDTA	PromoCell
Tween 20	Merck

3.1.2 Buffers and solutions

Designation	Source/ Composition
BCA protein assay reagent A	Th. Geyer Hamburg
BCA reagent B	4% (w/v) $CuSO_4$ * H_2O
Bromphenolblue solution	1% (w/v) Bromphenolblue in ethanol
BSA 1%	1% BSA in TBS-T
Developer solution	Kodack Medical
ECL solution I	2.5mM Luminol
	0.4mM p-Coumarsäure
	100mMTris-HCl (pH 8.5)
ECL solution II	100mM Tris-HCl (pH 8.5)
First strand buffer	Invitrogen/ Promega
Fixing solution	Kodack Medical
LB medium	10g/L Trypton
	5g/L yeast extract
	5g/L NaCl
	pH 7.5
PCR buffer (10x)	Q-Bio gene
RIPA buffer	1% NP 40 (100%) (Igepal CA 630)
	0.5% Sodium – deoxycholate
	0.1% SDS
	1mM EDTA
	1mM EGTA
	1mM Sodium orthovanadate
	20mM Sodium fluoride
	0.5mM DTT
Roti®-Load 1 (4x)	Roth
Rotiphorese® buffer (10x)	Roth
Rotiphorese Gel® 30 (37,5 : 1)	Roth
Separating gel buffer (pH 8.8)	0.4% SDS
	1.5 M Tris - HCl (pH 8.8)
Stacking gel buffer (pH 6.8)	0.4% SDS
	0.5 M Tris - HCl (pH 6.8)

Designation	Source
TAE buffer 10x	10mM EDTA (pH 8,0)
	2.3 % Essigsäure
	0.5 M Tris
TBS 10x	0.1 M Tris-HCl (pH 7.4)
	1.5 M Natriumchlorid
TBS-T	150mM Natriumchlorid
	10mM Tris - HCl (pH 7.4)
	0.1% Tween 20
TFB-1	100mM rubidium chloride
	50mM magnesium chloride
	30mM potassium acetate
	10mM calcium chloride
	15% glycerol
	pH 5.8
TFB-2	10mM MOPS
	100mM rubidium chloride
	10mM calcium chloride
	15% glycerol
	pH 6.8 adjusted with potassium hydroxide
	pH 5.8

3.1.3 Media for cell culture

Designation	Source
Dulbeccos MEM	PAN
OptiMem® Serum free medium	Gibco
SkMC Growth Medium Kit	PromoCell
SkMC Differentiation Medium Kit	PromoCell
Trypsin Neutralisation Solution	PromoCell
Fetal calf serum	Biochrom

MATERIAL AND METHOD

3.1.4 Oligonucleotides

3.1.4.1 Conventional PCR

table 2: Gene-specific primer used in convential PCR and for sequencing.
(fwd = forward-primer; rev = reverse-primer)

Gene name	NCBI-Accession Number	Oligo name	Primer sequences		Annealing temperature
CSNK1A1	NT_029289	CSNK1A1_Prom1	fwd	AAA TAC CTC AAC CGA CAA GG	
			rev	TCC TCT CAA TCT TCT TTG CAC	
		CSNK1A1_Prom2	fwd	GAG CGT GCA AAG AAG ATT G	
			rev	AAA ATG CCT TGC TGA CTC AC	
NFATc	NT_010879	NFATc_Prom1	fwd	GAC ACG AGT TTA GCC TTT GG	
			rev	CTG TGG AAA TTT GGA AAA GC	56°C
		NFATc_Prom2	fwd	TGG GAA TTT CCT TTC TAG GG	
			rev	TTT AAA GCT GGA AAA CAC CTC	
RAP2B	NT_005612	RAP2B_Prom1	fwd	ACT TAG GGG TGT CAG ACT CG	
			rev	ACC TGC CCC ATA GAT TTT TC	
		RAP2B_Prom2	fwd	AAT CTA TGG GGC AGG TTT G	
			rev	CTC TCT CCC ACA GCC TCT AC	

All gene-specific were designed with the computer program "Primer3" (http://www-genome.wi.mit.edu/cgi-bin/primer/primer3_www.cgi) and ordered from the company Biotez in Berlin, Germany.

The used primer concentration was 10µM at a time. In case of RAP2B, the sequence of 1kb upstream of the gene bore a GC-content 60-70%. This resulted in a use of a specific amplification procedure as described in section 3.3.9.

3.1.4.2 TaqMan PCR

table 3: Gene-specific TaqMan probes and primer used in TaqMan Real-Time-PCR. Probe and primer sequences of the assays on demand are not specified by the manufacturer (ABI).
(fwd = forward-primer; rev = reverse-primer; ns = not specified)

Gene name	Order number	Sequence of the probe	Primer sequence	
Calcineurin	Hs00174223_m1		n.s.	
c-Jun	Hs00277190_s1		n.s.	
CSNK1A1		CTT GTC GGT TGA GGT ATT T	fwd	GGT AGG CAG CTT CTT TGT TTC TTT
			rev	TGA TTT CTG GCT AAA AGT ATC AAA GGG TTT
JNK1	Hs00177083_m1		n.s.	
MYF5	Hs00271574_m1		n.s.	
NFATc		CAA GCC GAA TTC TC	fwd	GGG AGA TGG AAG CGA AAA CTG A
			rev	AAA TGG CGG GAT CTC AAC CA
p21		CTT CGG CCC AGT GGA CAG	fwd	ACC CAT GCG GCA GCA A
			rev	CGC CAT TAG CGC ATC ACA
RAP2B	Hs00267932_s1			
UTRNA		ATT GTG TTC ATC CAG ATC TG	fwd	AAT GGG CAG AAC GAA TTC AGT GAT A
			rev	CAT TTG GTA AAG GTT TTC TTC TGT ACG T
18s rRNA	4308329		n.s.	

Primers and probes ordered as assay on demand are indicated by their order numbers. All others were designed by Applied Biosystems (ABI). In both cases they were delivered as a ready to use mixture. The sequence of each mRNA including information of the position of exon-exon junctions was send to the company using the ABI software "Assay by design File builder" (version 2.0). For that purpose, the whole sequence of each gene and exon-exon junctions were verified using the NCBI Evidence Viewer in "LocusLink" (http://www.ncbi.nlm.nih.gov/LocusLink/) and the Ensembl Human Genome Server (http://www.ensembl.org/). All probes were designed to overlap an exon-exon junction and to hybridise between both primers. This characteristic eliminates the possibility of detection of nonspecific amplification and also of DNA contamination.

3.1.4.3 LightCycler PCR

table 4: Gene-specific primer used in LightCycler Real-Time-PCR; (cds = coding sequence; fwd = forward-primer; rev = reverse-primer)

Gene name	NCBI accession number	Oligo name	Primer sequences		Annealing temperature
CSNK1A1	NM_001892	CSNK1A1_cds270	fwd	GGA TCT TCT GGG ACC TAG CC	
		CSNK1A1_cds515	rev	TGT TGC CTT GTC CTG TTG TC	
GAPDH	NM_002046		fwd	GAA GGT GAA GGT CGG AGT C	56°C
			rev	GAA GAT GGT GAT GGG ATT TC	
GFP	FM883229.1	GFP	fwd	TCA GGA GGA CGA GAA ACA TC	
			rev	AAA GGG CAG ATT GTG TGG AC	
At DNA level					
CSNK1A1	NM_001892 NT_029289	CSNK1A1 ORF/XmnI	fwd	GTT CGA ACC ATG GCG AGT AG	
			rev	CCT TTC ATG TTA CTC TTG GTT TTG	
		CSNK1A1_Ex3	fwd	GGA TCT TCT GGG ACC TAG CC	56°C
		CSNK1A1_Int4_176	rev	CCA GGT TTT GTC TTA TTC CCA TAG	
ß-Actin	NM_001101 NC_000007	Act.3	fwd	TCT ACA ATG AGC TGC GTG TGG CTC	
		Act.4	rev	GCT CGG TGA GGA TCT TCA TGA GGT	

The GAPDH primer were designed by TIB MOLBIOL GmbH (Berlin, Germany). All other primer pairs were ordered from Biotez (Berlin, Germany).

3.1.5 Kits

Designation	Source
Advantage GC 2 PCR	Clontech
BrdU cell proliferation test	Roche
CCK8 cell vitality test	Dojindo

MATERIAL AND METHOD

Designation	Source
FaststartStart DNA^{Plus} SYBR Green I Master Mix	Roche
	Ambion
mirVana RNA prep Kit	QIAGEN
QIAprep Plasmid Maxi Kit EndoFree	QIAGEN
QIAquick Gel Extraction Kit	QIAGEN
QuantiTect reverse transcription Kit	QIAGEN
Plasmid Maxi Kit	Promega
RNase free DNase set	Invitek
Spin Cell RNA Mini Kit	Applied Biosystems
TaqMan 18s rRNA standard	Invitek
Twin Spin Cell Mini Kit	Roche
X-tremeGene siRNA Dicer kit	

3.1.6 Molecular weight marker and plasmid vector DNA

Designation	Source
HiMark^{TM} pre-stained protein standard	Invitrogen
Marker VIII	Roche
λ-Hind III	self made
ΦX174/HAEIII	self made
pReceiver M03 CSNK1A1	Genecopoeia
TrueClone expression vector CSNK1A1	OriGene

3.1.7 Enzymes

Designation	Source
AmpliTaq	Applied Biosystems
AmpliTaq Gold	Applied Biosystems
DNase I (Amp. Grade)	Roche

Designation	Source
HaeIII – enzyme + buffer	Boehringer I
RNase H	Invitrogen
RNase Out	Invitrogen/ Promega
SuperscriptII RT	Invitrogen
EcoRI – enzyme + buffer	MBI Fermentas
SpeI – enzyme + buffer	MBI Fermentas
StuI – enzyme + buffer	MBI Fermentas

3.1.8 Antibodies

Designation	Source
Anti ß-actin (human, mouse) monoclonal	Santa Cruz
POD-Anti-mouse IgG	Oncogene Research
Anti UTRNA (human, rat) monoclonal	Santa Cruz

3.1.9 Transfection reagents and systems

Designation	Source
amaxa NHDF nucleofection kit	amaxa
Effectene	QIAGEN
Fugene 6	Roche
Fugene HD	Roche
Lipofectamine 2000	Invitrogen
Targefect/ Virofect	Targeting system
X-tremeGENE siRNA transfection reagent	Roche

3.1.10 Equipment

Designation	Type	Source
ABI Prism SDS	7000	Applied Biosystems
Analytical balance	CP64	Sartorius
Cooling centrifuge	Sigma-3MK	Sigma
Econo Spin		Sorvall Instruments
DNA-Sequencer	3100	Applied Biosystems
Homogeniser tapered, Teflon®	358133 series	Wheaton Science
Horizontal gel electrophoresis	Horizon 11-14	Life Technologies
iBlot blotting system	Version C	Invitrogen
Image Master	VDS	Pharmacia Biotech
Incubator	BR6000	Heraeus
Laminar flow	HLB2472	Heraeus Instruments
LightCycler®	2.0	Roche
LightCycler® Rotor	2.0	Genaxxon
Magnetic stirrer	MR 3001K	Heidolph
Microplate reader	VersaMax	Molecular Devices
Micro quartz cuvette	50µL volume	Varian
Microscope		Zeiss Jena
Multichannel-pipet	2-200µL	Brand
PCR-Cycler	T3	Biometra
Pipettes	Research	Eppendorf
Pipet boys	Pipetus®-akku	Hirschmann
pH meter	CG 820	Schott
Photometer	Carry 100	Varian
Precision balance	CP2202S	Sartorius
Shaker Stovall	Belly Dancer	Castle Scientific
Standard power pack	P25	Biometra
Thermomixer	Compact	Eppendorf
Precision balance	Excellence	Sartorius
Table top centrifuge	Mini Spin	Eppendorf
Vortex	Genie 2	Bender
Waterbath	WB10	Memmert

3.1.11 Consumable supplies

Designation	Source
Cryo vials	Corning
Falcon tubes 15mL; 50mL	Corning
iBlot™ Gel Transfer Stack, Nitrocellulose	Invitrogen
LightCycler plastic capillaries	Genaxxon
Microscope slides	Roth
Neubauer counting chamber	Labor Optik
NuPAGE Mini Gels Tris-Acetate 4-12%	Invitrogen
Optical caps, 8 stripes	Applied Biosystems
Pasteur pipettes	Roth
Pipet tips (0.5µL – 1000µL)	Faust
Reagent reservoirs	Costar
Safe-lock tubes (0.5mL, 1.5mL)	Roth
8 stripes (nuclease free)	GeNunc
TC flasks 25cm^2; 75cm^2	TPP
Tubes (0.2mL - 2.0mL)	Roth
UV-cuvette	Eppendorf
X-ray films	Kodack Medical
6well cell culture plates	TRP
96well cell culture plates	TRP
96well Optical plates	Applied Biosystems

3.2 Software

table 5: Software and web interfaces for sequencing analysis and database searches

Name	Field of application	Version / Internet address
ABI PrismSDS	Software for analysis of Real-Time-PCR results	Version 1.2x
BioDocAnalyzer	Densitometrical analysis tool for Western blot results	Version 2.0 Biometra
Ensembl Human	Human genome, BLAST sequences	http://www.ensembl.org/human/
File builder	Sequence checker and text editor to build, edit and correct imported sequence submission files	Version 2.0 Applied Biosystems (Assay by Design Service)
GeneCards	Overview about human genes summarised from different databases	http://www.genecards.org
Human Genome Project working draft at UCSC	Gene maps, Alignment and search tool	http://genome.cse.ucsc.edu/index.html
KEGG database at Biocarta	Signal transduction pathway database and tool	http://www.biocarta.com
NCBI	Homologies, BLAST, Cytogenetic map, Genetic map, Genetic markers	http://www.ncbi.nlm.nih.gov/
Primer3	Primer design	http://frodo.wi.mit.edu/cgi-bin/primer3/primer3_www.cgi
OMIM	catalog of human genes and genetic disorders	http://www.ncbi.nlm.nih.gov/entrez/query.fcgi?db=OMIM
Sequencher	Sequencing analysis and alignment tool	Sequencher Demo Version 4.0.5 Gene Codes Corporation
SwissProt	Protein database	http://us.expasy.org/sprot/

3.3 Methods

3.3.1 Flow diagram

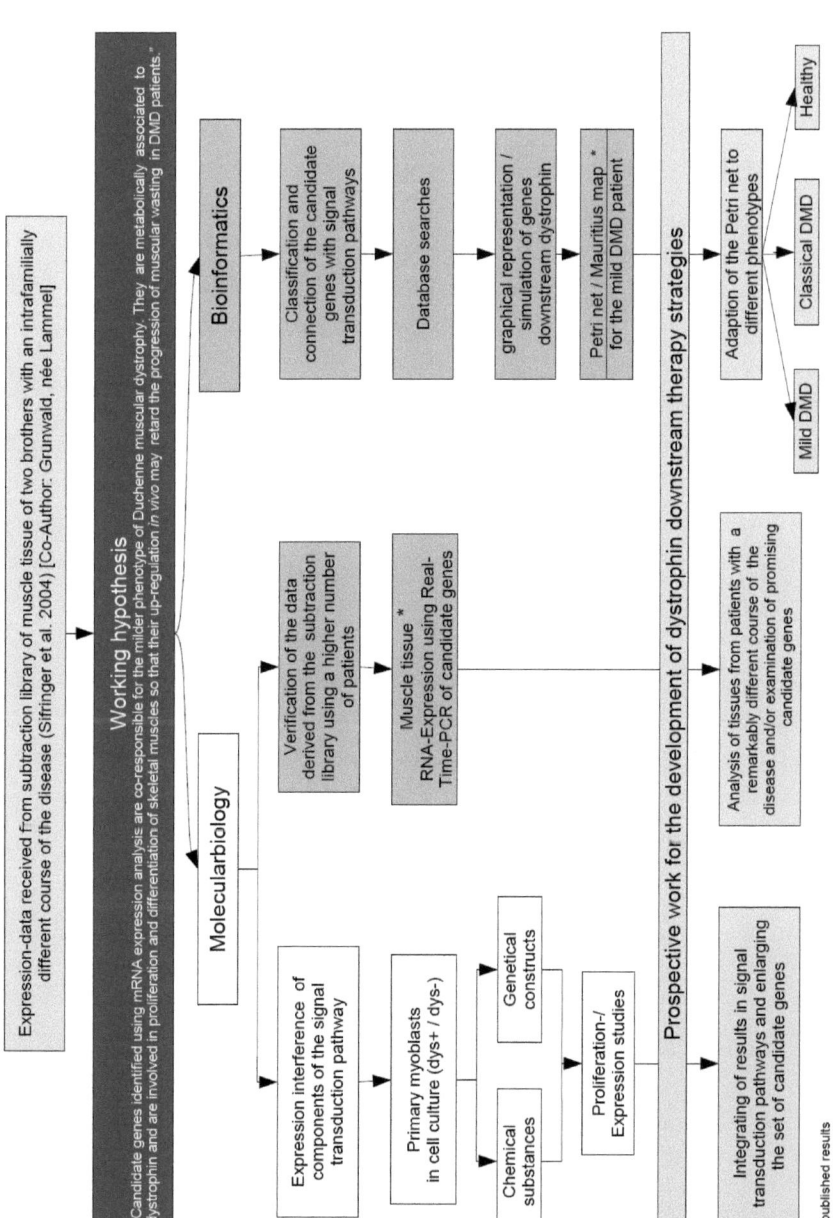

* published results

3.3.2 Origination of muscle tissue and human skeletal muscle cells

3.3.2.1 Muscle biopsy tissue

After informed consent according to the declaration of Helsinki muscle biopsies were usually taken from M. quadriceps femoris for diagnostic reasons. One section of the samples was immediately frozen in liquid nitrogen for immunohistochemical diagnostic as well as research and another one was used to generate a cell culture.

Muscle tissue was obtained by biopsy from DMD patients and healthy controls, and kindly provided by Dr. med. Arpad von Moers (Charité Berlin, Campus Virchow Clinic), frozen in liquid nitrogen and stored at $-80°C$. DMD patients as well as healthy controls were at different ages at the time of biopsy.
The clinical features and data of molecular diagnosis of the 2 affected brothers were described in Sifringer et al. 2004. In short, especially differences in sitting (18 months for the elder one versus 9 months for the younger) and walking (40 versus 18 months), and loss of gait (9 versus 12 years) make clear that at comparable ages the elder brother is much more affected by muscle weakness and remarkable progression of symptoms than the younger one. Because the elder brother died at age 16.5 years and muscle material was used for the subtraction library (Sifringer *et al.*, 2004) it was not available for further studies. Five DMD patients were used as equivalent DMD patients. Patient p1 presents comparable clinical features similar to the elder brother. He started walking at 30 months and lost gait at 10 years. The time point of muscle biopsy of both siblings was age-matched at age 6.
In addition, another 4 DMD patients were analysed to confirm the outstanding position of the younger brother. As normal controls, 5 specimens of diagnostic biopsies were used that turned out to be normal. The age at biopsy of the controls varies between 1 year, 7 months and 11 years and for the classic DMD-patients between 3 years, 5 months and 8 years.
An overview of clinical features and molecular diagnosis data of 5 classic DMD patients as well as of the brothers is given in table 6.

MATERIALS AND METHODS

table 6: Clinical and molecular data of healthy controls and DMD patients who are donors of muscle biopsy tissues used in mRNA expression analyses (section 3.3.7)

	Age at biopsy	Muscle type	Diagnosis	Walking	Wheelchair-bound
			Healthy controls		
c1	11 years	M.Quadriceps femoris	osteochondrom		
c2	28 months	M.Quadriceps femoris	(intermit.) CK level slightly increased	n.d.	
c3	19 months	M.Quadriceps femoris	ponto-cerebell. hypoplasy; epilepsy; microcephaly		
c4	33 months	M.Quadriceps femoris	microdeletion 7q36		
c5	3 years 10 months	M.Quadriceps femoris	mental and physical retardation; (intermit.) increased CK level; microcephaly		
			DMD patients		
pm	6 years	M.Quadriceps femoris	immunhistological dystrophin negative	18 months	12 years
ps	6 years	M.Quadriceps femoris	immunhistological dystrophin negative	40 months	9 years
p1	7 years 5 months	M.Bizips femoris	immunhistological dystrophin negative	30 months	10 years
p2*	2 years	M.Quadriceps femoris	immunhistological dystrophin negative	16 months	walks (4 years)
p3	8 years	M.Quadriceps femoris	immunhistological dystrophin negative	15 months	walks (9 years)
p4	7 years 5 months	M.Quadriceps femoris	deletion of exon 7-39 immunhistological dystrophin negative	18 months	10 years
p5*	3 years 5 months	M.Quadriceps femoris	immunhistological dystrophin negative	12 months	walks (6 years)

c = controls; p = classic DMD patient; pm = mild DMD patient; ps = elder brother of pm
n.d. = not determined
For patients pm, ps, p1, p2, p3 and p5 no mutations were found at DNA level. But due to clinical signs and immunohistological dystrophin negative data there is no doubt concerning the diagnosis of DMD.
* in cell culture table 7

3.3.2.2 Primary human skeletal muscle cells (myoblasts) used in cell culture

Primary human skeletal muscle cells (SkMCs) were received from biopsy tissue of the right or left quadriceps of healthy controls and DMD as well as BMD patients (Dr. med. Arpad von Moers, Charité Berlin, Campus Virchow Clinic, Berlin, Germany) and prepared at the MTCC ("Muscle Tissue Culture Collection", Munich, Germany).

Selection of myoblasts from the biopsy samples was performed according to an established procedure of the MTCC, Munich, Germany. To obtain primary human myoblast cultures, muscle tissue samples were treated proteolytically. Separation of myoblasts from other cell types contained in muscle tissue was carried out in the group of Prof. Dr. Hanns Lochmueller by Eva Neugebauer using the MACS method. The magnetic cell separation technique through magnetic beads with the use of cell-specific antigens on the surface is very efficient. Vitality and purity of the obtained cells are very high (Dickson *et al.*, 1987; Webster *et al.*, 1988; Walsh *et al.*, 1989; Walsh, 1990).

The separated cells were cryo-preserved for transportation from the MTCC Munich to Berlin. After arrival the cells were first cultured for 2 passages before they were frozen in liquid nitrogen again (see section 3.3.3). An overview of the primary cells used in subsequent experiments is shown in table 7.

table 7: Overview origin of biological samples

Designation		Clinical data			Cell culture	
		Classification	Age at biopsy	Dystrophy	Doubling of population	Subjective impression cultured cells
14/00		normal	2 years	-	4d	normal
18/01		normal	9 years	-	5.5d	normal
43/01		DMD	3 months	no impairment	6d	slightly delayed
77/02		DMD	4 years	progressive, mild course	12d	delayed; myoblasts enlarged
109/03	P5*	DMD	3.5 years	progressive	12d	delayed; myoblasts enlarged
145/03	P2*	DMD	2 years	progressive	7d	slightly delayed

* patients also available for muscle biopsy investigations in table 6

3.3.3 Culturing of primary human skeletal muscle cells (myoblasts)

Culture conditions

SkMCs were cultured under humid conditions at 37°C and 5% (V/V) carbon dioxide (CO_2) in an incubator (Heraeus Instruments). The culture media containing 5% fetal calf serum (FCS) and growth factors (PromoCell) was changed every 3 to 4 days at a final volume of 200µl/cm² depending on the proliferation rate of the cells. The growth of the cells was examined microscopally (Zeiss, Jena) every 24 hours. After the monolayer reached a sub-confluence of 80-90%, the cells were passaged. A washing step with 200µl/cm² of 1 x PBS (Biochrom) was performed followed by trypsinisation of the cells with 50µl/cm² Trypsin/EDTA solution (PromoCell). The serine protease trypsin hydrolyses the peptide bonds and therefore dislodges adherent cells from the walls of culture flaks. This can be advanced by gently tapping the side of the culture flasks. When all cells were detached, trypsinisation was stopped using 100µl/cm trypsin neutralisation solution (TNS, PromoCell). The cells were harvested and transferred into a 15mL falcon tube (Corning) followed by centrifugation at room temperature for 5min at 200 x g (Econo Spin). The supernatant was removed without disturbing the cell pellet. The cell cluster was re-suspended in fresh growth medium and distributed according to their proliferation rate into new culture flasks.

Long-time storage

For long-time storage as well as transportation of the primary human myoblasts, the cells were cryo-preserved in Cryo-serum free medium (Cryo-SFM; PromoCell) containing DMSO, methyl-cellulose and other cryo-protectant ingredients. Myoblasts were harvested as described above but re-suspended in cold Cryo-SFM at a concentration of 1-5 x 10^6 cells per 1mL of Cryo-SFM, transferred into a cold cryo tube (Corning) and gradually frozen to −196°C for final storage in liquid nitrogen (Thermo N_2 container). For thawing the cryo vial was gently shaken in warm water (37°C) until most of the Cryo-SFM was thawed. The cell suspension was transferred into a 15mL falcon tube with 10mL pre-warmed growth medium. Remaining cell suspension in the cryo vial was thawed with warmed medium and also added to the falcon tube. After centrifugation as described above, the supernatant was removed and the cell pellet was re-suspended in warm growth medium at a concentration of 2 x 10^5 cells per 1mL. The cell solution was displaced in a new pre-warmed culture flask.

The sizes of the culture flasks and plates are dependent on the cell number distributed to the culture vessel. An overview of cell numbers applied to different culture containers is given in table 8.

table 8: The number of primary human skeletal muscle cells applied to used culture flasks and plates.

Culture vessel	Cell Growth Area	Cell number
96well plate	0,32cm²	3 x 10^3
6well plate	9cm²	3-4 x 10^5
T25 flask	25cm²	1 x 10^6
T75 flask	75 cm²	3 x 10^6
T150 flask	150 cm²	6 x 10^6

C2C12 myoblasts and HeLa cells were cultured as described for primary human skeletal muscle cells, but in Dulbecco's Modified Eagle's Medium (DMEM) including 10% FCS instead of SkMC growth medium.

3.3.4 Transfection

Various transfection reagents and systems are available. But numerous primary cell types, such as the primary human skeletal muscle cells, are very hard to transfect and resist most of them. Therefore, 4 different transfection reagents as well as a electroporation system were tested to discover the optimal transfection conditions in terms of vitality and transfection efficiency: Effectene™ (QIAGEN), Fugene 6® (Roche), Fugene HD® (Roche), Targefect® (Targeting System) and the amaxa nucleofection system (amaxa). In the case of Fugene HD, the vector DNA was not only transfected into primary human myoblasts as for the other transfection reagents, but also into HeLa and C2C12 cells to examine whether the insert of the vector DNA is differentially expressed in different cell types.

Plasmid vector DNA

Two different types of expression clones with a cDNA insert encoding CSNK1A1 were used in transfections: TrueClone™ (Origene) and OmicsLink™ ORF Expression Clones (Genecopoeia).
The OmicsLink™ ORF Expression pReceiver M03 is a full length protein coding ORF clone and contains the cDNA insert (from start to stop codon), 2 antibiotic cassettes (neomycin, ampicillin) and the green fluorescent protein (GFP) as reporter gene. Both, GFP and the insert are regulated through the same cytomegalovirus promoter (CMV) at which translation of GFP is initiated by the internal ribosome entry site (IRES). The reporter gene GFP allows the observation of successful transfection in a fluorescence microscope without any isolation of DNA or RNA. Neomycin can be used as a selection marker in mammalian cells to induce a stable transfection. The whole plasmid is about 7kbp in size.
The Origene TrueClone™ pCMV6-XL5 has the designated full length cDNA cloned downstream of the CMV and T7 promoter and contains the resistance gene for ampicillin. A reporter gene is not included. The length of the clone is 7.3kbp.
Before use, both vectors were transformed into E.coli to produce unlimited amounts of plasmid DNA.

Preparation of competent E.coli cells

For the preparation of competent E.coli cells, 100mL of LB-medium containing 50µg/mL ampicillin were inoculated with cells from the glycerol stock and incubated on a shaker at 37°C over-night. Next day, 1mL of the cells was transferred into 100mL pre-warmed LB-medium and again incubated on a shaker (100rpm) at 37°C until the optical density (OD600) of 0.5 was reached. A volume of 50mL was moved to a centrifuged tube and left on ice for 5min. After centrifugation at 4°C and 4000 x g for 5min, all the following steps were performed on ice. The supernatant was discarded and the cell cluster was re-suspended in 15mL pre-chilled buffer TFB-1 followed by another incubation for 90min on ice. After a second centrifugation step at 4°C and 4000 x g for 5min, the cells were gently re-suspended in 2mL pre-chilled buffer TFB-2. The cell suspension was distributed to 1.5mL reaction tubes with a volume of 100µL per tube and stored at –80°C.

Transformation of DNA in E.coli

The transformation reaction was set up for chemical competent E.coli cells as described in table 9:

table 9: Reaction set-up for transformation of plasmid DNA into competent E.coli cells

Reagent	Volume [µL]
Competent E-coli	100µL
Vector DNA (100ng/µL)	2µL
LB-Medium	500µL
Final volume	**602µL**

The reaction containing E.coli and vector DNA was mixed gently and incubated for 15-20min on ice. Afterwards, the cells were heat-shocked for 45sec at 42°C without shaking and then immediately transferred on ice for 2min. To the reaction, 500µL of to 37°C pre-warmed LB medium was added. The tubes were placed on a shaker at 37°C for 2 hours. The total volume of each transformation was centrifuged at room temperature for 5min at 3000 x g, re-suspended in 100µL LB-medium and then spread on pre-warmed selective plate containing ampicillin and incubated at 37°C over night. Next day, a colony was picked and the plasmid DNA was isolated using the QIAGEN Plasmid Mini kit. The plasmid DNA was analysed by PCR and gel electrophoresis to confirm the presence and the correct length of the insert. For transfection, a high amount of plasmid DNA was required. For that purpose, 100mL of LB-medium containing 50µg/mL ampicillin was inoculated with cells from original colony and incubated at 37°C over-night. Next morning, 1mL of the cell suspension was transferred into 250mL fresh LB-medium and incubated again at 37°C for 4-6h. Cells were harvested and the plasmid DNA was isolated using the QIAGEN Plasmid Endotoxin Free Maxi Kit. The whole procedure was performed as described in the manual of the kit. The obtained plasmid DNA was measured using OD260 and used for further transfection.

For long-term storage, the original analysed colony was inoculated into 10mL of LB containing 50µg/mL ampicillin and the culture was grown over-night. Next day, an aliquot of this culture was applied to a solution of 50% LB-medium and 50% sterile glycerol. The glycerol stock was stored at -80°C.

Transfection using amaxa nucleofection system (amaxa)

The amaxa nucleofection system is based on the electroporation technique and was already applied successfully for primary human SkMCs by Quenneville et al 2004 (Quenneville *et al.*, 2004). As suggested by the authors, the human dermal fibroblast (NHDF) nucleofector kit was used for transfection (Quenneville *et al.*, 2004). The whole procedure was applied as described in the manufacturer's instruction. Different programs were investigated, but only one at a time (table 10).

table 10: Preparation and programs used for nucleofection for primary human myoblasts

Primary human myoblasts	NHDF solution	Vector DNA (max. 5µL)				Program			
0.5-1.0 x 10^6	100µL	2µg	4µg	6µg	8µg	X01	P20	U20	U25

Pre-warmed growth medium was added to the cuvette and the cell solution was distributed to 3 35mm culture dishes. The third dish also contained a sterile cover slip used for fluorescence microscopy. The cells were cultured in the incubator at 37°C, 5% CO_2, and 95% humid atmosphere. Negative controls treated either with DNA or the nucleofection reagent alone were always performed at the same time. Next day, the medium was replaced with medium containing the selection marker neomycin and cover slip was prepared to determine transfection efficiency using fluorescence microscopy (see below). The transfection efficiency was also investigated by determination of copy numbers and expression rate of plasmid DNA using Real-Time PCR (as described below).

The nucleofection system was not utilised in combination with Origene TrueClone™ pCMV6-XL5 vector.

Transfection using Effectene™ (QIAGEN)

Effectene™ is a non-liposomal lipid, which is used in combination with a DNA-condensing enhancer and optimised buffer to reach high transfection efficiencies. Firstly, the enhancer condenses the DNA molecules and the Effectene reagent coats them with cationic lipids. Thus, this allows a transfection of primary eukaryotic cells.

Transfection was applied in 6-well plates with normal primary human myoblasts according to manufacturer's instructions. Effectene™ was already utilised in transfection of antisense oligonucleotides in this group. Therefore, the procedure was performed with the same protocol (table 11). After 6h of culturing in the incubator at 37°C, 5% CO_2, and 95% humid atmosphere, the medium was replaced.

table 11: Effectene™ transfection conditions performed with 4 x 10^5 primary human myoblasts

Buffer EC	DNA (in 20µL)	Enhancer	Effectene™
80µL	1µg 5µg	3.2µL	20µL

Negative controls treated either with DNA or the transfection reagent alone were simultaneously performed. Two days after transfection, the achievement of the transfection was monitored using fluorescence microscopy. As mentioned for transfection with amaxa nucleofection system, the transfection efficiency was also investigated by determination of copy numbers and expression rate of plasmid DNA using Real-Time PCR.

Effectene™ was not utilised in combination with Origene TrueClone™ pCMV6-XL5 vector.

Transfection using Fugene 6® and Fugene HD® (Roche)

Fugene 6® as well as Fugene HD® are non-liposomal lipids supplied with other components in 80% ethanol and suitable for both, stable and transient transfections. The protocol of the kit was followed at all times but with various ratios of Fugene : DNA as suggested by the protocol. Both, Fugene 6® and Fugene HD® are low in toxicity. In the case of Fugene HD, transfection was applied to primary human myoblasts, HeLa and C2C12 cells, to examine whether the insert of the vector DNA is differentially expressed in different cell types.

For transfection, the cells were trypsinised and plated into 6-well plates. The number of cells used per well is shown in table 12. Contrary to the Fugene HD® protocol, the transfection was applied directly after seeding of the cells. In the case of Fugene 6®, the transfection took place the next day as suggested by the manual, and sterile cover slips were included in the

MATERIAL AND METHOD

6-well plates. The following procedure as described for Fugene HD® below was applied for Fugene 6® in the same way.

table 12: Number of cells for each cell type as used in transfection with Fugene HD® in 6-well plates

Cell type	Cell number
C2C12 myoblasts	5×10^5
HeLa	1×10^6
primary human myoblasts	4×10^5

The Fugene HD® transfection was performed using different Fugene HD : DNA ratios as well as different volumes of the Fugene HD®-DNA-complex solution as suggested by the protocol. The procedure is published in Grunwald and Speer (Grunwald and Speer, 2007).

table 13: Different transfection parameters as tested for primary human myoblasts transfection efficiency

Ratio Fugene HD® (µL) / DNA (µg)	Negative control				Fugene HD® : DNA		
	DNA only	Fugene HD® only					
	0:2	3:0	6:0	12:0	3:2	6:2	12:4
Volume of complex (%)	100% 200%	100% 200%	200%	200%	100% 200% 400%	100% 200% 400%	100% 200% 400%

In the case of Fugene 6®, transfection efficiency was observed using a fluorescence microscope. The Origene TrueClone™ pCMV6-XL5 vector was not applied in transfection with Fugene 6®, but with Fugene HD®. After 72h of transfection with Fugene HD®, the cells were washed, trypsinised and washed again with 1x PBS before the lysis buffer was added. Transfection efficiency was determined by examination of plasmid copy numbers and expression rate using Real-Time PCR. Negative controls as shown in table 13 were performed in each experiment.

A second transfection with Fugene HD® was performed in order to investigate its influence on proliferation and vitality of the cells, 72h after transfection. Since the proliferation test has to be applied in 96-well plates but transfection is more efficient in 6-well plates, the cells were transferred from 6-well plates to 96-well plates after 24h of transfection.

Materials and Methods

Transfection using Targefect/Virofect® (Targeting Systems)

Targefect F-2 is a non-lipid cationic polymer with DNA-condensing properties and is low in toxicity. Targefect F-2 is used in combination with Virofect. Virofect is an adenovirus-derived formulation, which has the ability to complex with plasmid DNA via the cationic targefect F-2. It can enhance gene transfer by increasing cellular uptake and allows the escape of the transfected plasmid DNA from lysosomal degradation because of its ability to lyse the endosome. But it does not contain any replication-competent virus.

Plasmid DNA was diluted in serum free OptiMem®, mixed by vortexing before and after addition of targefect F-2 (table 14). Virofect was included, the solution was mixed, and left for 20min at 37°C to form complexes. Afterwards, the transfection complex was placed onto the cells and the dish was swirled to mix the transfection complex with growth medium. The cells were incubated for 24h at 37°C, 5% CO_2 and 95% humid atmosphere until the medium was changed with an additional wash step using fresh medium, next day.

table 14: Conditions for transfection with Targefect/ Virofect using 4×10^5 for primary human myoblasts

Condition	OptiMem SFM	DNA	Targefect F-2	Virofect
1	0.6mL	6µg	12µL	-
2	0.6mL	6µg	12µL	25µL
3	1.2mL	3µg	9µL	22.5µL

Negative controls treated either with DNA or the transfection reagent alone were performed at the same time. Two days after transfection, the transfection efficiency was determined using fluorescence microscopy.

Targefect/Virofect was not utilised in combination with Origene TrueClone™ pCMV6-XL5 vector.

Detection of successful transfection using a fluorescence microscope

The green fluorescent protein GFP is a protein found in the jellyfish aequorea victoria. It fluoresces green when it is exposed to blue light. GFP is widely used as a reporter gene in DNA constructs as proof of transfection.

After 48h of transfection, the cover slips included in the 6-well plates were first washed twice with 1 x PBS followed by fixation of the cells with 3,7% formaldehyde in 1 x PBS for 1h. The cover slips were washed again, briefly dried and covered with ProLong® Antifade (Invitrogen) mounting medium on a microscope slide. The GFP fluorescence was detected using a fluorescence microscope with a filter of 480nm.

MATERIAL AND METHOD

Detection of successful transfection using Real-Time-PCR (LightCycler)
For absolute quantification of transfected plasmid vector copy numbers the Real-Time-PCR was used (see section 3.3.9). Therefore molecular weight of both plasmid DNA constructs were calculated to determine the plasmid copy numbers per ng. The concentration of pure plasmid DNA was determined by measurement of optical density (OD260) for both vectors to establish a standard curve ranging from 0.05ng to 8ng. These standard curves were used to recalculate the plasmid DNA copy numbers in transfected cells.

Preparation and transfection of siRNA
For transfection of siRNA, esiWay Silencing Resources® (RZPD, Berlin, Germany) was applied. The so-called esiRNA uses pools of siRNA molecules (esiRNA) to eliminate screening of multiple synthetic siRNAs for an effective knockdown. As specified by the manufacturer RZPD, this technique is supposed to reduce off-target effects due to very low concentrations of individual siRNA molecules, compared to synthetic siRNAs.

The sequences were delivered as purified PCR products containing 2 opposing T7 promoter sequences. X-tremeGENE siRNA Dicer Kit (Roche) was used to prepare ready-to-transfect d-siRNA. The instructions of the manufacturer were followed strictly during set-up and procedure. The concentration of siRNA was determined by OD260.

Transfection of siRNA into primary human SKMCs was employed by using the X-tremeGENE siRNA transfection reagent (Roche), a lipid-based transfection reagent that forms a complex with siRNA. The instructions were followed at all times with minimal changes and with various concentrations of siRNA in initial experiments.

For transfection, the cells were trypsinised and 4×10^5 cells were plated into 6-well plates. The transfection was performed using 3 different amounts of siRNA and a volume of 10µL of X-tremeGENE siRNA transfection reagent as suggested by the protocol (table 15). At first, 1µg of siRNA was diluted in 100µL of serum and antibiotic free medium (OptiMem, Invitrogen). In a separate tube, 10µL of the transfection reagent was added to 90µL OptiMem medium and transferred into the diluted siRNA followed by an incubation time of 15-20min. This mixture was added drop-wise to the cells at once (=200µL). In contrast to the protocol, the transfection complex was applied directly after seeding of the cells. The cells were incubated and harvested after 48h and 72h, respectively. Negative controls were performed as shown table 15. For negative controls, the transferred volume was added to 200µL with OptiMem.

MATERIALS AND METHODS

table 15: Different transfection parameters as tested for primary human myoblasts transfection efficiency

	Untreated	Negative controls		X-tremeGene siRNA®: siRNA		
		siRNA only	X-tremeGene siRNA® only			
Ratio X-tremeGene siRNA® (µL) / siRNA (µg)	0:0	0:1	10:0	3:2	6:2	12:4
Added Volume	200µL	200µL	200µL	200µL	200µL	200µL

A second transfection was performed in order to investigate its influence on proliferation and vitality of the cells after 48h and 72h of transfection. Since a proliferation test has to be applied in 96-well plates but transfection is more efficient in 6-well plates, the cells were transferred from 6-well plates to 96-well plates after 24h of transfection.

The success of transfection was investigated by Real-Time-PCR after 48h and 72h of transfection. After trypsinisation and an additional washing step with 1x PBS, RNA was isolated using the Spin Cell RNA Mini Kit (Invitek) according to manufacturers' instructions. The concentration and purity of the RNA was determined by the OD260 and the ratio to OD280, respectively. Reverse transcription was performed using QuantiTect® reverse transcription kit (Qiagen) in conformity with the directions. The kit provides reverse transcription of RNA and removal of genomic DNA in one step. The gene knockdown was examined by relative quantification as described in section 3.3.9.

3.3.5 Treatment of primary human myoblasts with chemical substances

The primary human myoblasts were treated with different chemical substances in order to investigate their influence on proliferation and vitality of the cells. Additionally, the mRNA expression of 3 genes, MYF5, p21 and UTRNA, was examined in treated cells. The compounds were chosen according to their effect on proteins such as phosphatases as drawn in the signal transduction pathway (figure 11a-d in section 4.1). Nonetheless, the fact needs to be taken into consideration that chemical compounds affect a wide range of target proteins.

For proliferation and vitality studies, the primary human myoblasts were seeded into 96-well plates with 3×10^3 cells per well. The reagents were directly applied to myoblasts in supension prior to seeding of the cells. Alterations in proliferation (BrdU) and vitality (WST-8) of the cells were examined after 24h, 48h, 72h and in some cases also after 96h. For longer incubation than 24h, the medium was changed every 24h to fresh medium including chemical substances. As negative controls, untreated cells as well as cells treated with the

MATERIAL AND METHOD

solvent of the compounds were involved in each experiment. The table 16 shows the chemical substances used their solvents and the final concentration of the solvent in the cell medium in addition to the site of pharmacological effects within the cells.

table 16: Chemical compounds tested in primary human myoblasts

Chemical substance	Solvent	Site of the pharmacological effects
CsA	DMSO (7mM)	Antagonist of calcineurin (Sharma et al., 1993)
Deflazacort	DMSO (7mM/ 70mM)	Stimulant of proliferation (Biggar et al., 2006)
IC261	DMSO (70mM/ 7mM)	Inhibition of casein kinase I (Knippschild et al., 2005b)
Okadaic acid	DMSO (7mM)	Phosphatase inhibitor (Rodova et al., 2004)

Cyclosporine A, Deflazacort, IC261 and okadaic acid were dissolved in 14M dimethoxysulfoxide (DMSO). Because of the toxicity of DMSO, the first dilution step (1:10) was performed using pure skeletal muscle growth medium, not only to dilute the chemical compound but also to dilute DMSO down to 1.4M (except for CsA as described below). All the following dilution steps were carried out with 1.4M DMSO suspended in growth medium. A final concentration of 7mM DMSO was applied to the cells.

Cyclosporine A

The immunosuppressive drug CsA is widely used in organ transplantation. It leads to a decrease of the activity of the immune system and prevents organ rejection. CsA binds to cyclophilin resulting in a complex formation, which leads to inhibition of calcineurin and inactivation of calcium dependent activation of the cell, especially T-lymphocytes.

At high molar concentrations (>10mM), CsA is not soluble in 1.4M DMSO. Hence in experiments with final CsA concentrations ranging between 1-100μM the stock solutions were diluted in 14M DMSO resulting in a final concentration of 70mM in the respective experiment. All other concentrations were supplied in 1.4M DMSO. A review of the applied concentration in each experiment is given in table 17.

MATERIALS AND METHODS

table 17: Concentration of CsA used in different primary cells at several time points and the final concentration of its solvent DMSO

Cells		Range of final concentrations used in the respective experiment	Time points
(Dys+)	18/01	1-50nM CsA (7mM DMSO) 1-100µM CsA (70mM DMSO)	24h 48h 72h 96h
	43/01		
(Dys-)	77/02		
	109/03		
	145/03		

Deflazacort

Deflazacort is a glycocorticoid. It is used to reduce inflammation and allergic responses such as asthma. Thus, similarly to CsA, it can also prevent the rejection of organ transplants.

Deflazacort is not soluble in physiological saline and 1.4M DMSO at high molar concentrations (>1mM).

The range of concentrations of deflazacort used in different primary cells is shown in table 18. The solvent of deflazacort acid and the procedure of solving are described at the beginning of this section 3.3.5.

table 18: Concentration of deflazacort used in different primary cells at several time points and the final concentration of its solvent DMSO

Cells		Range of final concentrations used in the respective experiment	Time points
(Dys+)	14/00	1µM – 1mM deflazacort (7mM DMSO) 10mM – 200mM deflazacort (70mM DMSO)	48h 72h
	18/01		
(Dys-)	43/01		
	77/02		
	109/03		
	145/03		

IC261

IC261 (3-[(2,4,6-trimethoxyphenyl)methylidenyl]-indolin-2-one) was selected because of its ability to inhibit all isoforms of casein kinase 1. Members of this protein family are CSNK1 epsilon, CSNK1 delta and CSNK1 alpha. It has been suggested that CSNK1 delta and CSNK1 epsilon are involved in DNA repair and chromosomal segregation. The protein CSNK1A1 inactivates the transcription factor NFATc, whereby expression of p21, a negative regulator of the cell proliferation, may be reduced. IC261 inhibits CK1d

MATERIAL AND METHOD

(IC50 = 0.7 - 1.3 mM), CK1e (IC50 = 0.6 - 1.4 mM) and CK1a1 at much higher concentrations (IC50 = 11 - 21 mM).

In table 19, the range of concentrations of IC261 used in different primary cells is shown. The solvent of IC261 and the procedure of solving are described at the beginning of this section 3.3.5.

table 19: Concentration of IC261 used in different primary cells at several time points and the final concentration of its solvent DMSO

Cells		Range of final concentrations used in the respective experiment	Time points
(Dys+)	14/00		24h
		1-100nM IC261 (7mM DMSO)	
	43/01		48h
(Dys-)		1-100µM IC261 (7mM DMSO)	
	77/02		72h

Okadaic acid

Okadaic acid is a toxin which was originally isolated from the sponge Halichondria okadai. The okadaic acid sold on the market is produced by the dinoflagellate Prorocentrum. It potently inhibits serine/threonine phosphatases such as calcineurin.

The range of concentrations of okadaic acid used in different primary cells is shown in table 20. The solvent of okadaic acid and the procedure of solving are described at the beginning of this section 3.3.5.

table 20: Concentration of okadaic acid used in different primary cells at several time points and the final concentration of its solvent DMSO. [OA = okadaic acid]

Cells		Range of final concentrations used in the respective experiment	Time points
(Dys+)	18/01	0.1-1nM OA (7mM DMSO)	24h
	43/01	1-100nM OA (7mM DMSO)	48h
(Dys-)	77/02	1-100µM OA (70mM DMSO)	72h

3.3.6 Cell Proliferation and vitality tests

Proliferation and vitality of the primary human myoblasts were determined after transfection or treatment with chemical substances. Both tests were performed in 96-well plates with 100µL cell growth medium per well and 24h after treatment of the cells with chemical substances and 48h after transfection at the earliest.

Cell proliferation assay using BrdU (Roche)
Proliferation of the cells correlates directly with DNA synthesis. The 5-bromo-2'-deoxyuridine (BrdU)-assay is a colorimetric immunoassay for quantification of cell proliferation, based on the measurement of BrdU incorporation during DNA synthesis. The protocol of the kit was followed at all times and all reagents used were provided by the kit. The colour development was photometrically measured after 10min, 20min and 30min using an ELISA reader at 370 nm (reference wavelength: 492 nm).

Cell vitality assay using WST-8 (Dojindo)
The test is based on the intra-cellular reduction of the tetrazolium salt WST-8 (2-(2-methoxy-4-nitrophenyl)-3-(4-nitrophenyl)-5-(2,4-disulfophenyl)-2H-tetrazolium, monosodium salt) by cellular de-hydrogenases to an orange formazan. Formazan is soluble in cell culture medium and the amount is directly proportional to the number of living cells. The assay was performed by adding 10µl of the WST-8 solution directly to the cells. Approximately 2h after incubation at 37°C, 5% CO_2 and 95% humid atmosphere, the absorbance of the samples was measured in a microplate reader at 450nm (reference wavelength: 650nm).

3.3.7 Isolation of DNA and RNA

RNA was isolated from muscle tissue and from primary human skeletal muscle cells by TriReagent (Sigma) and Spin Cell RNA Mini Kit (Invitek)/Twin Spin Cell Mini Kit (Invitek), respectively. DNA and protein isolation was performed for primary human skeletal muscle cells. Highly purified plasmid DNA was received from transformed E.coli using the EndoFree Plasmid Maxi Kit (QIAGEN).

Plasmid DNA isolation using EndoFree Plasmid Maxi Kit (QIAGEN)
For transfection of primary human myoblasts a higher amount of plasmid was required. Transformation of E.coli with plasmid DNA is described in section 3.3.4. The plasmid DNA was isolated using the EndoFree Plasmid Maxi Kit (QIAGEN) to receive highly purified plasmid DNA. The whole procedure was performed as described in the manual of the kit. DNA concentration was determined using OD260.

Genomic DNA isolation using Twin Spin Cell Mini Kit (Invitek)

For absolute quantification of transfected plasmid copy numbers, DNA was isolated from transfected cells using Twin Spin Cell Mini Kit according to manufacturer's instructions. DNA concentration was determined using OD260.

RNA isolation using TriReagent® (Sigma)

The disruption of cells in culture and muscle tissue was performed in different ways.

Muscle tissue

The frozen muscle tissue sample were cut and transferred into a mortar including 1mL of TriReagent®. The tissue was homogenised using a tapered tissue grinder with a Teflon® pestle. The subsequent procedure was performed as described for cells from cell culture below.

Cells

Cells dissolved in 1mL of TriReagent® were vortexed and stored for 5min at room temperature to ensure complete lysis of the cells. Afterwards a phenol-chloroform extraction and ethanol precipitation were applied. For that purpose, 200µL chloroform was added to the homogenate. After vortexing for 15sec and incubation for 15min at room temperature, the samples were centrifuged for 15min at 12.000 x g and 4°C. The aqueous phase was transferred into a new 1.5mL tube containing 0.5mL of isopropanol. The samples were mixed by inversion, incubated for 30-60min at −20°C and centrifuged for 20min at 12.000 x g and 4°C. The supernatant was carefully discarded and the pellet was washed twice with 500µL of 70% ethanol. The ethanol was removed completely and the pellet was dried under air conditions before it was dissolved in 30-50µL DEPC water.

The RNA was treated with DNase I to receive DNA-free total RNA.

RNA isolation using Spin Cell RNA Mini Kit or TwinSpin Cell Mini Kit (Invitek)

For RNA-extraction from primary human myoblasts, Spin Cell RNA Mini kit was primarily applied. In case of RNA isolation from transfected cells, the TwinSpin Cell Mini kit was used for simultaneous DNA/RNA isolation from the same cells. The very similar protocols of both kits were followed at all times and all used buffers and spin filters were provided by the kit except for ethanol and ß-mercapto ethanol. The RNA was treated with DNase I to receive DNA-free total RNA.

Treatment of RNA with DNase I

To obtain DNA-free total RNA, incubation with DNaseI was required. For that purpose, a reaction mix was prepared as shown in table 21. RNA isolated from transfected or chemical substance treated cells was reverse transcribed by the QuantiTect reverse transcription kit, which already includes a step to remove DNA beforehand and was proving the DNaseI treatment unnecessary.

table 21: DNase I reaction mix for removing DNA in RNA samples

Compound	Volume
DNaseI nix	10µL
RNA dissolved in water	25µL
Water	55µL
10U/µL DnaseI (pure enzyme)	10µL
Σ	100µL

The reaction mix was added to each RNA sample and incubated at 37°C for 20min. The reaction was stopped by a denaturation step at 95°C for 5min. Afterwards a phenol-chloroform extraction and ethanol precipitation were used to purify DNA-free RNA. For that purpose, 100µL of DEPC water and 200µL of Roti phenol/chloroform/isoamylalcohol (25:24:1) were added to the RNA. After vortexing for 10sec, the samples were centrifuged for 10min at 12.000 x g. The aqueous phase was transferred into a new 1.5mL tube containing one volume of chloroform/isoamylalcohol (24:1). The tubes were vortexed for 30sec followed by another centrifuge step. Again, the aqueous phase was transferred into a new 1.5mL tube containing 1/10 volume of 3M sodium acetate (pH 5.2), 1.0µL of glycogen and 3 volumes of absolute ethanol. The samples were mixed by inversion of the tubes and stored at –20°C for 1-2h followed by centrifugation for 30min at 12.000 x g and 4°C. The supernatant was carefully discarded and the pellet was washed twice with 200µL of chilled 70% ethanol. The ethanol was removed completely and the pellet was dried under air conditions before it was dissolved in 20-50µL DEPC water.

The concentration of the RNA was determined using OD260. The RNA was stored at –80°C for long term.

3.3.8 Reverse transcription of DNA-free RNA

The small amount of RNA obtained from muscle tissue was reverse transcribed using random priming in 0.5ml tubes by the SuperScript II reverse transcriptase (Invitrogen). The SuperScript II enzyme has a reduced RNase H activity, which consequently eliminates degradation of RNA molecules during first-strand cDNA synthesis and leads to a greater yield of full-length cDNA. The reaction was performed under the following conditions (table 22):

table 22: Reverse transcription of RNA from muscle tissue using SuperScript II RT kit (Invitrogen)

Compound	Volume [µL] or Quantity [µg]
DNA free total RNA	500ng to 1.0µg
10mM dNTP's (Invitrogen)	1.0µL
50ng/µL pd(N)$_6$ (Invitrogen)	2.0µL
Nuclease-free water (QIAGEN)	To a final volume of 16.0µL
Incubation: for 5min at 70°C tubes on ice	
5x 1st strand buffer (Invitrogen)	5.0µL
0.1M DTT (Invitrogen)	2.5µL
rRNasin – RNase inhibitor (40U/µL – Invitrogen)	0.5µL
Incubation: for 2min at 42°C	
SuperScript II RT (200U/µL – Invitrogen)	1.0µL
Incubation: for 60min at 42°C	
Deactivation: for 15min at 70°C	
Storage: forever at -20°C	
Σ	25.0µL

In later experiments, the RNA from human SkMCs was reverse transcribed under the use of Quantitect reverse transcription kit. This kit provides efficient removal of residual DNA and reverse transcription in one step and eliminates the loss of RNA through the purification steps of RNA after DNA digestion. The reaction was performed as suggested by the manufacturer table 23.

table 23: Reverse transcription of RNA from SkMCs using Quantitect reverse transcription kit (Qiagen)

Compound	Volume [µL] or Quantity [µg]
total RNA	500ng to 1.0µg
gDNA Wipeout Buffer 7x	2.0µL
Nuclease free water (QIAGEN)	to a final volume of 14µL
Incubation: for 4min at 42°C	
afterwards tubes on ice	
Quantiscript reverse transcriptase	1.0µL
5x Quantiscript RT buffer	4.0µL
RT primer mix	1.0µL
Incubation: for 15min at 42°C	
Deactivation: for 3min at 95°C	
Storage: forever at -20°C	
Σ	20.0µL

3.3.9 PCR

PCR is an acronym, which stands for polymerase chain reaction. The PCR technique is basically a primer extension reaction for amplifying specific nucleic acids *in vitro*.

Conventional PCR

A conventional PCR was performed using standard PCR condition. All utilised PCR compounds were manufactured by Q-Bio gene. The samples were prepared as follows (table 24):

table 24: Preparation of reaction mix for conventional PCR

Compound	Volume [µL]
ddH$_2$O	8.5µL
25mM MgCl$_2$ (2,5mM)	5.0µL
10x PCR buffer (1x)	2.0µL
4mM dNTP-mix (250µM)	2.0µL
10µM gene-specific forward primer (1µM)	0.5µL
10µM gene-specific reverse Primer (1µM)	0.5µL
Taq Pol (5U/µL)	0.5µL
1st strand cDNA (100ng)	1.0µL
Σ	20.0µL

MATERIAL AND METHOD

At first a master mix was prepared by combining all the components except the cDNA and subsequently sub-divided into 0.2mL PCR tubes into equal parts. Afterwards cDNA (~100ng) was added, the tubes were mixed briefly and the reaction was performed in the PCR machine from Biometra immediately using the following settings:

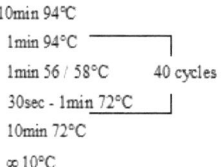

As negative controls samples with water and not reverse-transcribed RNA instead of template (cDNA) were used to exclude DNA contamination in any of the compounds. To check the success of any PCR, a gel-electrophoresis was performed each time. For that purpose 2µL of 10x DNA-loading buffer was added to 10µL of PCR-product and loaded into each well of a 1% agarose gel in 1x TAE-buffer. A molecular-weight marker was also used for fragment sizing. Electrophoresis was usually performed for 60min at 80V. Afterwards, the agarose gel was stained with ethidium bromide, visualised on a UV transilluminator, and documented.

The 1kb upstream sequence of RAP2B presented a high GC content (60-70%) which arose in difficulties to amplify this region. The PCR amplification kit "Advantage GC 2 PCR" from Clontech was be successfully applied in this procedure. The kit was utilised as described by the manufacturers.

Real-Time-PCR using LightCycler

Real-Time-PCR analysis using LightCycler was performed to measure the amount of transfected vector DNA. Additionally, RNase H and the restriction enzyme StuI, which cuts the plasmid once, were added to the isolated DNA to remove RNA contamination and to linearise the circular plasmid DNA. SYBR Green I was used as fluorescence stain instead of oligonucleotide probes. The Real-Time-PCR reaction was prepared according to manufacturers' instructions of the FastStart DNA Master^{Plus} SYBR-Green I mix (Roche) (table 25). The protocol was used as published in Grunwald and Speer (Grunwald and Speer, 2007).

MATERIALS AND METHODS

table 25: Preparation of Real-Time PCR reaction using FastStart DNA Master[Plus] SYBR-Green I

	Endogenous DNA	Exogenous DNA
Forward primer (10µM)	0.5µL	
Reverse primer (10µM)	0.5µL	
FastStart DNA Master[Plus] SYBR-Green I	4µL	
Water	10µL	
DNA	5µL (50ng)	5µL (1ng)
Σ	20µL	20µL

Real-Time PCR using TaqMan

For Real-Time PCR the ABI Prism 7000 Sequence Detector from Applied Biosystems was used. Theoretically, a quantitative relationship exists between the starting amount of the target sequence and the amount of the product at the end of any given PCR cycle. That means the higher the fluorescence signal and the earlier it is detectable the higher the numbers of initial copy number. TaqMan is a hydrolysation probe relying on fluorescence resonance energy transfer (FRET) for quantification. The protocol and the calculation procedure were performed as published in Sifringer et al. (Sifringer *et al.*, 2004)

For relative quantification, 18S rRNA was used as an internal standard but prepared for each cDNA separate to the other samples. For each sample, 3 replicates were prepared to prove the equability of each reaction.

The Real-Time PCR was performed using the TaqMan Universal Master Mix and MicroAmp Optical plates and caps. The following pipetting scheme was performed (table 26):

table 26: Real-Time PCR reaction using TaqMan

Component	Volume [µL]
20X working stock of Gene Expression Assay Mix	1.0
cDNA template + RNase-free water	9.0
2x TaqMan Universal Master Mix	10.0
Σ	20.0

For each analysed gene, a water control was included. The real time mix was prepared and an aliquot was applied to each well of a 96-well plate. The plate was briefly centrifuged to avoid air bubbles in the reaction mix. The real time PCR was performed using the following conditions:

MATERIAL AND METHOD

```
10min 50°C
10min 94°C
1min 94°C  ┐
           │ 40 cycles
1min 60°C  ┘
```

The data were analysed using the appropriate TaqMan Real-Time PCR software from Applied Biosystems.

3.3.10   Sequencing

For DNA-Sequencing the chain termination method of Sanger was used. The cycle Sequencing was performed with ABI PRISM® BigDye™ Terminators v3.0 Cycle Sequencing Kit using a different protocol to Applied Biosystems. The labelled DNA was detected through ABI PRISM® 3100 Genetic Analyser. The ABI PRISM® 3100 Genetic Analyser is an automated, multi-colour fluorescence-based DNA analysis system using the technology of capillary electrophoresis. Samples are prepared in 96-well plates.

Preparing the reactions

First PCR was performed as described in 0, the whole PCR volume was loaded on a 1% agarose gel and the DNA bands were cut under UV-light. Clean and purified DNA was obtained using the QIAGEN gel extraction kit. Yields of DNA following cleanup were determined by Agarose gel analysis. The amount of sample DNA loaded was estimated by visual comparison of the band intensity with that of the standards of the λ-HindIII-ladder. Template quantities of 1-3ng per 100bp of PCR-products were required for the BigDye™ terminator chemistry for a 1x cycle sequencing run. For each reaction, the following reagents were added to a separate 0.2µL tube (table 27):

table 27: Composition of sequencing reaction

Reagent	*Quantity*
Terminator Ready Reaction Mix	2.5µL
PCR product DNA	1-3ng per 100bp (max. 100ng)
10µM Primer (fwd or rev)	1.0µL
Deionised water	ad to 10µL
Σ	**10.0µL**

The tubes were mixed well and spun briefly followed by performing the cycle sequencing under the conditions below:

```
1min    96°C
20sec   96°C   ⎤
10sec   *°C     30 Cycles
4min    60°C   ⎦
∞       10°C
                * primer-specific annealing temperature
```

These settings differ from the standard protocol suggested by Applied Biosystems.
In the next step unincorporated dye terminator must be completely removed before the samples can be analysed. For this purpose an ethanol precipitation of the extension products was performed. The tubes were removed from the cycle sequencing run and to each tube 8µL of deionised water and 32µL of 96% ethanol were added. The tubes were left at room temperature for 15min to precipitate the extension products followed by centrifuge for 45min at 3200 x g. The supernatant was discarded by inverting the tubes onto a paper towel. A volume of 150µL of 70% ethanol was added to each well to rinse the pellet followed by another centrifuge step for 15min at 3200 x g. The supernatant was discarded again and the pellets were left to dry under air conditions for at least 10min. The pellets were dissolved in 20µL milli-Q water for 30min at room temperature placed on a shaker. Finally, the solutions were delivered to the Institute of Medical Molecular Diagnostics GmbH (IMMD, Berlin) where the sequencing process was performed. The sequence data were analysed using Sequencher demo version 4.0.5 (Gene Codes Corporation).

3.3.11 Protein analysis

Protein isolation
A cell number of 1×10^6 was lysed in 200µL RIPA buffer by repeated up- and down-pipetting of the solution followed by centrifugation for 10min at 3000 x g and 4°C. The supernatant was centrifuged again for 20min at 17,000 x g and 4°C to obtain the cytosolic protein fraction. The samples were stored at –80°C for long term.

Quantification of proteins

The bicinchoninic acid (BCA) assay was applied for protein quantification in 96-well plates. The BCA assay is based on the biuret reagent and included a series of known standard concentrations of the protein bovine serum albumin (0-200µg/mL). Both, the samples and the standard proteins were performed in duplicates. For determination, a volume of 200µL of the BCA reagent combined with 1M copper sulphate (50:1) was added to 50µl of the sample or standard followed by an incubation at 50°C for 30min. The optical density was investigated at 550nm using a micro-titre-plate reader. Finally, the protein concentration was examined by linear regression of the known concentration of the standard protein.

Western blot

SDS-PAGE (sodium dodecyl sulfate poly-acryl-amide gel electrophoresis) is a widely used technique for Western blots.

Because of the difference in molecular weight of (UTRN) and the standard protein ß-actin, the Western blot analyses was performed using a gradient Tris acetate mini gel (4-12%). For determination of UTRN, 40µg of the protein sample was applied. The protein samples were mixed with 2x NuPAGE protein sample buffer, 1µl antioxidant and 10x reducing agent (table 28), and were incubated at 85°C for 5min. The proteins are so prepared for separating their polypeptide chains according to their molecular weight by PAGE (Laemmli, 1970) and transferring to an immobilised membrane (blotting) for detection with specific antibodies.

table 28: Preparation of protein samples for SDS-PAGE and Western blot

Component	Volume
Protein sample (40µg)	xµL
2x NuPAGE sample buffer	20µL
NuPAGE Antioxidant	1µL
10x Reducing agent	4µL
DEPC-dH$_2$O	xµL
Σ	**40µL**

The HiMark^{TM} pre-stained high molecular weight protein standard was included in each SDS-PAGE. The electrophoresis was carried out at 125V for 100min. The proteins contained in the SDS-gel were transferred onto the surface of a nitrocellulose membrane (iBlot™ gel transfer stack, nitrocellulose) using the iBlot™ dry blotting system. The protein transfer was accomplished using the program P3 (20V) for 13min. The pre-stained protein standard was visible on the membrane and indicated a protein transfer performed. To verify the successful transfer of the protein samples, the membrane was stained for 30sec with Ponceau S (Roth) followed by 2 washing steps in TBS-T for 10min to remove the stain. The nitrocellulose membrane was blocked using 5% skim milk in TBS-T over-night. Next day, the membrane was washed 4 times in TBS-T for 10min and incubated in an antibody solution (1% BSA/TBS-T) including the antibody UTRN (1:100) or ß-actin (1:50000) over-night. Afterwards, the membrane was washed again 4 times in TBS-T for 10min followed by incubation with the second peroxidase-bound antibody solution (mouse IgG; 1:2000 in 1% BSA/TBS-T) for 120min. Another wash step followed as described above. The next steps were performed in the dark under red light. For incubation with the substrate, $1.8\mu l$ H_2O_2 was added to 3mL of ECL-solution II, mixed briefly, transferred to 3mL ECL-solution I and thoroughly mixed again. The combination was spread onto the membrane surface and incubated for 60sec. The membrane was removed and placed into a film cassette and covered with an X-ray film. The exposition time was continued for approximately 120min and 3sec for UTRNA and ß-actin, respectively. Developing of the X-ray film was performed by the successive use of a developer solution, a water bath and a fixing dilution. Subsequently, the protein signals were densitometrically analysed and quantified using the BioDocAnalyzer 2.0.

For the detection of the second protein, the standard protein ß-actin, the membrane was washed twice in TBS-T for 10min and the incubated in stripping buffer for 30min at room temperature to remove the bound antibodies of the first detection (UTRN). Afterwards, the membrane was washed again 4 times in TBS-T for 10min followed by blocking the membrane using 5% skim milk in TBS-T (Roth) over-night. All subsequent steps were performed as described above.

3.4 Bioinformatics tools

The graphical Petri net editor and animator Snoopy (Fieber, 2004) was applied for modelling the Petri net. Clustering was done with PINA (Grafahrend-Belau, 2006). Both tools are available free via http://www-dssz.informatik.tu-cottbus.de/swwwdssz/. The Integrated Net Analyzer – INA (Starke and Roch, 1999) was used for the analysis of structural and dynamic properties. For computation and representation of the Mauritius map we used an as yet unpublished tool of our own (Ackermann, 2008).

4 Results

Outlier patients with abnormalities in genotype/phenotype correlations, in particular patients experiencing an unexpected mild course of DMD, suggest modifications in gene expression and/or signal transduction pathways partially compensating the dystrophin defect. The identification of appropriate candidates which might be responsible for this phenomenon is one main basis for dystrophin downstream therapy strategies. In this group a first analysis of the transcriptom of such rare mild cases revealed 6 over-expressed cDNA clones (Sifringer *et al.*, 2004). However, neither connections to dystrophin nor relations between them were obvious. The same fact was found in other studies investigating the transcriptoms of DMD patients (Bakay *et al.*, 2002; Haslett *et al.*, 2002; Noguchi *et al.*, 2003; Haslett *et al.*, 2003; Baker *et al.*, 2006). But in further experiments such knowledge is absolutely crucial for establishing strategies to slow down the dystrophic process or, more ambitiously to restore muscle-function. Therefore the first part of this thesis is the identification of such connections by the use of experimentally gained data, as mentioned above. Additionally, these findings are supplemented by the results of the determination of p21 levels in DMD patients (Endesfelder *et al.*, 2000), scientific publications and data bases.

4.1 Signal transduction pathways connecting genes which were found differentially expressed in transcriptoms of selected DMD patients

The 3 genes, casein kinase 1 alpha 1 (CSNK1A1), ras related protein 2B (RAP2B), and p21, were connected to signal transduction pathways involving the transcription factor NFATc as a key component, figure 11a.

The resulting schematic network is grouped into 3 sections:
   (1) the phosphorylation pathway of NFATc
   (2) the de-phosphorylation pathway of NFATc
   (3) and the expression of target genes of NFATc

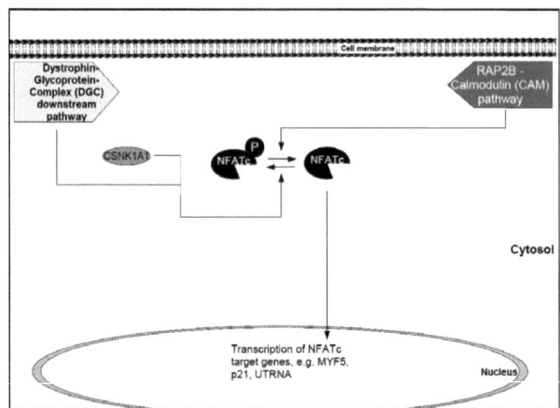

figure 11a: Outline of the biological processes.
The pathways are grouped into 3 parts. On the left side, the dystrophinglycoprotein complex (DGC) downstream cascade, which phosphorylates and de-activates NFATc. The de-phosphorylation and activation part of NFATc via RAP2B-calcineurin signalling is shown on the right side. The transcription factor NFATc mediates gene expression of several target genes, e.g. MYF5, p21, UTRNA.

The above-mentioned 3 parts are described in figure 11b-d in more detail.

The phosphorylation pathway (figure 11b) connects the DGC with NFATc and c-Jun via kinases such as c-Jun NH2-terminal kinase-p46 (JNK1) (Chow et al., 2000) and CSNK1A1. In the presence of a functional DGC, binding of laminin recruits RAC1 via syntrophins followed by phosphorylation of JNK1 and the transcription factor c-Jun itself (Oak et al., 2003). Phosphorylated c-Jun is capable of inhibiting the transcription of p21 followed by an increase in proliferation as demonstrated for C2C12 cells (Schreiber et al., 1987; Shaulian et al., 2000). For de-phosphorylation of c-Jun, only few experimental data are available. For example, LPS is described as a reagent to induce de-phosphorylation of c-Jun (Morton et al., 2003; Yu and Shah, 2004).

JNK1 as well as CSNK1A1 have the ability to phosphorylate the transcription factor NFATc (Schulz and Yutzey, 2004). Phosphorylated NFATc is inactive and located within the cytosol of the cell (Hogan et al., 2003).

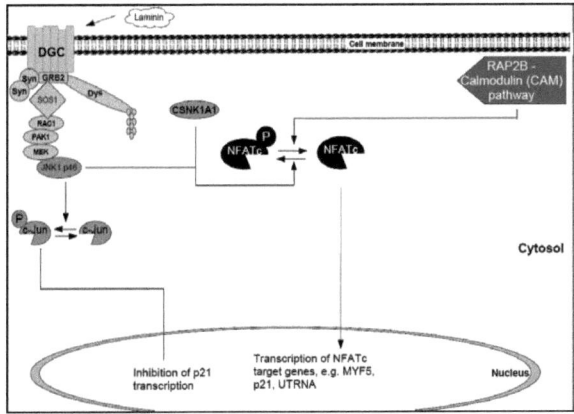

figure 11b: DGC downstream cascade leading to activation of JNK1 and c-Jun. JNK1 or CSNK1A1 are able to phosphorylate and de-activate NFATc.

In the de-phosphorylation pathway (figure 11c) of NFATc, RAP2B and calcineurin are principal components bridging this signal transduction cascade (Chin et al., 1998; Schmidt et al., 2001). Phospholipase C epsilon (PLCe) is activated by RAP2B (Keiper et al., 2004) followed by the release of calcium from the endoplasmatic reticulum via inositoltrisphosphat (IP3) (Evellin et al., 2002; Groenendyk et al., 2004). Calmodulin binds free calcium and activates calcineurin, which then can de-phosphorylate NFATc (Crabtree and Olson, 2002; Groenendyk et al., 2004).

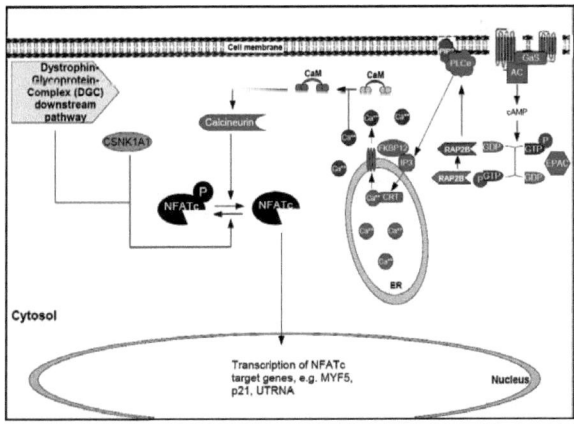

figure 11c: RAP2B-calcineurin-pathway. Active calcineurin mediates de-phosphorylation and activation of NFATc that lead to migration of NFATc into the nucleus.

De-phosphorylated NFATc migrates into the nucleus and acts as transcription factor for the expression of several target genes (figure 11d) (Macian, 2005), such as MYF5 (Friday and Pavlath, 2001), UTRNA (Chakkalakal *et al.*, 2003), and p21 (Santini *et al.*, 2001). Other examples of target genes are alpha actin (aActin), myosin (Hogan *et al.*, 2003), and atrial natriuretic factor (ANF), which, for clarity, are not included in figure 11d. In all the cases mentioned additional factors regulating these genes are involved. These genes form a complex with NFATc, e.g. SP1, SP3, GATA, GABP, MEF2 (Santini *et al.*, 2001).

The target gene UTRNA is homologous to dystrophin, and its function is described in section 1.3.2.1. MYF5 belongs to the myogenic regulatory factors family (MRF). MRFs are transcription factors specific to skeletal muscle cells. They regulate cell differentiation, and induce transcription of skeletal muscle-specific genes, e.g. creatine kinase, myosin light chains, and myosin heavy chains (MHC). MYF5 is one of the first genes expressed in differentiation events of myoblasts, and commits somatic cells to the myogenic lineage (Kosek *et al.*, 2006).

The protein p21 is not only regulated by NFATc but also by p53. It is also able to inhibit the cyclin-dependent kinases 2 and 4 (CDK2, CDK4) resulting in the diminishment of cell proliferation (Seoane *et al.*, 2002). The main target protein of CDK2 and CDK4 is the retinoblastoma protein (RB-protein). The de-phosphorylated RB-protein acts as transcriptional repressor by forming a complex with the E2F transcription factor whose function is required for the entry into cell proliferation (Zhang *et al.*, 2004). Sequential phosphorylation of RB by CDK4/CDK6 or CDK2 prevents the RB-protein from binding and inactivating of E2F, which results in the expression of S-phase genes (Harbour *et al.*, 1999).

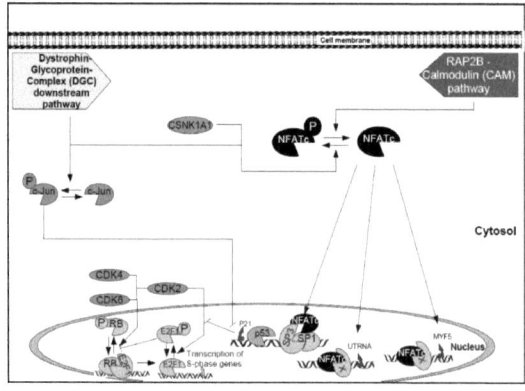

figure 11d: Expression of NFATc target genes. De-phosphorylation of NFATc by calcineurin unmasks the nuclear localization signal of NFAT, allowing its migration into the nucleus and initiation of specific genes carrying NFAT-responsive elements.

The signal transduction pathways as shown in figure 11b-d form the basis of the subsequent Petri net model. The comparison of the transcriptome of one mild and one typically severe DMD phenotype patient provided the foundation for these pathways (for clinical features see sections 4.2.1 and 4.4.1). In addition to modelling a Petri network, these experimental expression data were verified using an increased number of typically severe DMD patients and normal controls. Furthermore, not only CSNK1A1 and RAP2B but also other genes from the Petri net were analysed.

Comparison of the transcriptomes of a mild and a typically severe DMD phenotype patient provided the foundation for these pathways

4.2 Verification and extension of mRNA expression data

4.2.1 Muscle tissues originating from Duchenne muscular dystrophy patients and controls

In the first study, the very rare case of 2 brothers with an intra-familially different course of DMD was researched using the subtraction library technique. The clinical features and data of molecular diagnosis of the 2 affected brothers were described in (Sifringer *et al.*, 2004).

Because the elder brother died at 16.5 years and muscle material was used up for a subtraction library, an equivalent DMD patient (p1) as reference was applied for further investigations. Patient p1 exhibits clinical features similar to the elder brother. He started walking at 30 months and lost gait at 10 years. In addition, another 4 DMD patients were analysed to confirm the outstanding position of the younger brother. Muscle biopsies were obtained from classic DMD patients between aged 3 years 5 months and 8 years. For both brothers, biopsy was performed at age 6 years. As normal controls, 5 dystrophin positive specimens of diagnostic biopsies were used, which turned out to be normal.

An overview of clinical features and molecular diagnosis data of all patients and controls is given in section 3.3.2.1.

RESULTS

4.2.2 Expression analysis of selected genes at mRNA level

In addition to CSNK1A1, RAP2B, and p21, expression analysis included calcineurin, c-Jun, and NFATc, and its target genes MYF5 and UTRNA.
In, figure 12 mean values of expression data of 5 normal individuals, 5 DMD patients and the DMD with a mild course of the disease are shown. DMD patients differed in 4 genes from controls. They showed a significant increase at mRNA level for p21 and CSNK1A1 (P<0.01) compared to (Dys+) controls. Expression studies of RAP2B, UTRNA, and MYF5, also revealed an increase at mRNA level, which was not significant. In contrast, calcineurin as well as NFATc, JNK1, and c-Jun were not differentially expressed.
Expression level of patient pm noticeably differed from DMD patients. The results gave strong evidence that mRNA expression of p21 and CSNK1A1 were significantly decreased in patient pm compared to DMD patients (P<0.05), but unchanged compared to normal controls. Compared to the mean value of DMD patients, expression levels of RAP2B, NFATc, UTRNA, and MYF5 were reduced, but not significantly. The mRNA levels of calcineurin, JNK1, and c-Jun remained unaltered. In comparison to (Dys+) individuals, the expression levels of NFATc and JNK1 (P<0.05) were significantly decreased in patient pm, whereas calcineurin, UTRNA, and MYF5 were determined as differentially expressed.

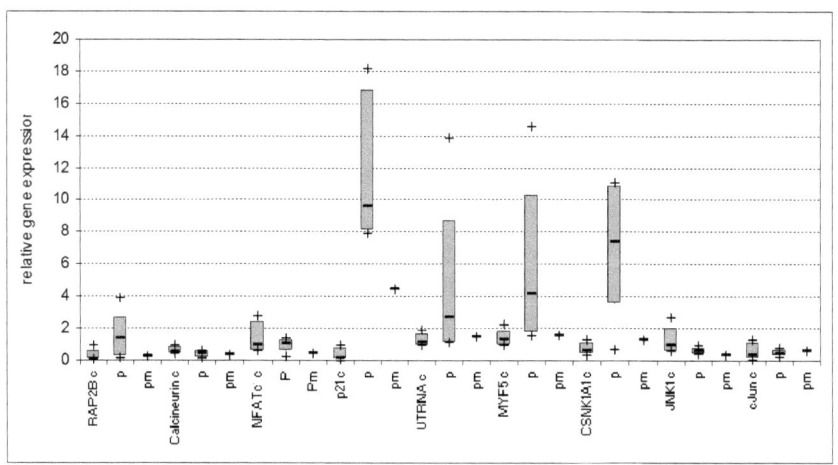

figure 12: Statistical analysis (students t-test, Box Whisker Plot) of relative mRNA expression data received from 8 genes involved in signal transduction network that is described in section 4.1. Skeletal muscle biopsy tissue of 5 DMD patients with typical course of disease and compared with 5 control individuals as well as a DMD patient presenting mild clinical findings. The median is indicated by a black rectangle. c = control; p = DMD patient; pm = mild DMD patient

In summary, differences in mRNA levels first determined in the 2 brothers with a varying course of the disease were partly also detected when comparing an increased number of DMD patients and controls. Interestingly, a milder phenotype seems to be associated with p21, CSNK1A1 and restricted with RAP2B expression. Whereas p21 was already subject of previous investigations (Endesfelder et al. 2000, 2003, 2005), CSNK1A1 and RAP2B are new candidates for interventions into pathways downstream dystrophin (section 4.4). However, the expression data of the 2 brothers were not confirmed, and especially the direction of the alteration of the CSNK1A1 expression level is controversial. The schematic network presented in section 4.1 provides basic fundamentals for the insight into a part of dystrophin downstream processes. However, it does not allow for modelling and incorporation of gene regulation and protein interactions. In contrast, Petri nets facilitate such modelling of gene regulations and thereby allow the simulation of pathological processes. Using Petri net theory the model could be checked for consistency by exploring the whole system behaviour. It opens up new potentialities to perform knockout experiments *in silico* for the determination of essential processes within the net. These results are described in the next chapter.

4.3 Systems biology

The objective of this work is modelling and analysis of a part of pathomechanisms of DMD in a Petri net. It is based on new experimentally gained data (4.2), literature data, and on the graphical representation of the dystrophin-enclosing signal transduction pathways presented in section 4.1.

4.3.1 The Petri net model

For the first time, modelling of a network connecting dystrophin, p21, CSNK1A1 and RAP2B was performed using Petri nets. The Petri net is depicted in figure 13. The net contains 64 places and 88 transitions. The complete list of both places and transitions including denotation is to be found in Grunwald et al. [Supplementary material (Grunwald *et al.*, 2008)].

Places represent genes, proteins, protein complexes, and complexes between proteins and nucleic acids. In the case of genes, a distinction was made between silencers and enhancers for gene regulation. Transitions stand for complex formation, binding, de-/activation, de-/phosphorylation, transcription etc. In addition to literature knowledge and experimentally obtained data, the network models some hypotheses regarding those processes which are not described in the literature and cannot be experimentally verified yet. Hypotheses have been introduced to fill gaps in the current knowledge to make the model consistent. For example, the location of NFATc phosphorylation by JNK1 or CSNK1A1 is not clearly known. Due to the fact that NFATc needs to be inactivated by phosphorylation and consequently to remigrate into the cytosol, the phosphorylation process mentioned above should also be feasible in the nucleus. Modelling this suggestion, transitions 81.deact_NFATc_CSNK1A1 and 82.deact_NFATc_JNK1 were introduced. Other examples concern the places 62.hypo_Silen_UTRNA and 63.hypo_Silen_MYF5, which represent the initiation of silencing processes after up-regulation of the transcription of the corresponding protein. Since in gene regulatory networks no stoichiometry is given, arc weights were in general set to one. Deviations from this rule were motivated by experimental data and the dynamic behaviour in simulations. According to experimental data of the patients, the arc weights were increased to indicate the delay of down-regulation in the following cases: t42.down_reg_UTRNA,

41.down_reg_MYF5, 28.down_reg_RAP2B, 53.down_reg_JNK1, 37.down_reg_CSNK1A1, and 84.bind NFATc_DNA_genes. Because of increased incoming arc weights of transitions, pre-conditions (i.e. the pre-places) had to carry more tokens to enable firing of transitions. For example in figure 14, the transition 53.down_reg_ JNK1 is enabled if there are at least 2 tokens on place JNK1. Thus, the transition 53.down_reg_JNK1 is delayed compared to the transitions 53.up_reg_JNK1 and act_JNK1. Arc weights in cases of protein interaction were also increased to reflect the experimental results. They are summarised in table 29.

table 29: The transitions describing protein interactions, which are indicated by a larger arc weight than one according to experimental data. For DGC, a special subnet structure has been chosen as exemplarily depicted in figure 14.

Protein interaction	Transition
DGC activation	t56
Phosphorylation of NFATc by CSNK1A1	t9
Phosphorylation of NFATc by JNK1	t60
Phosphorylation of RB	t65
Phosphorylation of E2F	t68

To avoid accumulation of tokens, arc weights were increased for t78.reg_p21_p53 and t87.degrad_calcineurin. Modelling of inhibitions into the Petri net was performed in different ways. A transition-induced inhibition was modelled as the removal of tokens from the system. For example, the kinase CDK2 was inhibited in the presence of p21, whereas E2F, phosphorylated by CDK2, inhibited the dissociation of the E2F_RB_compl.

RESULTS

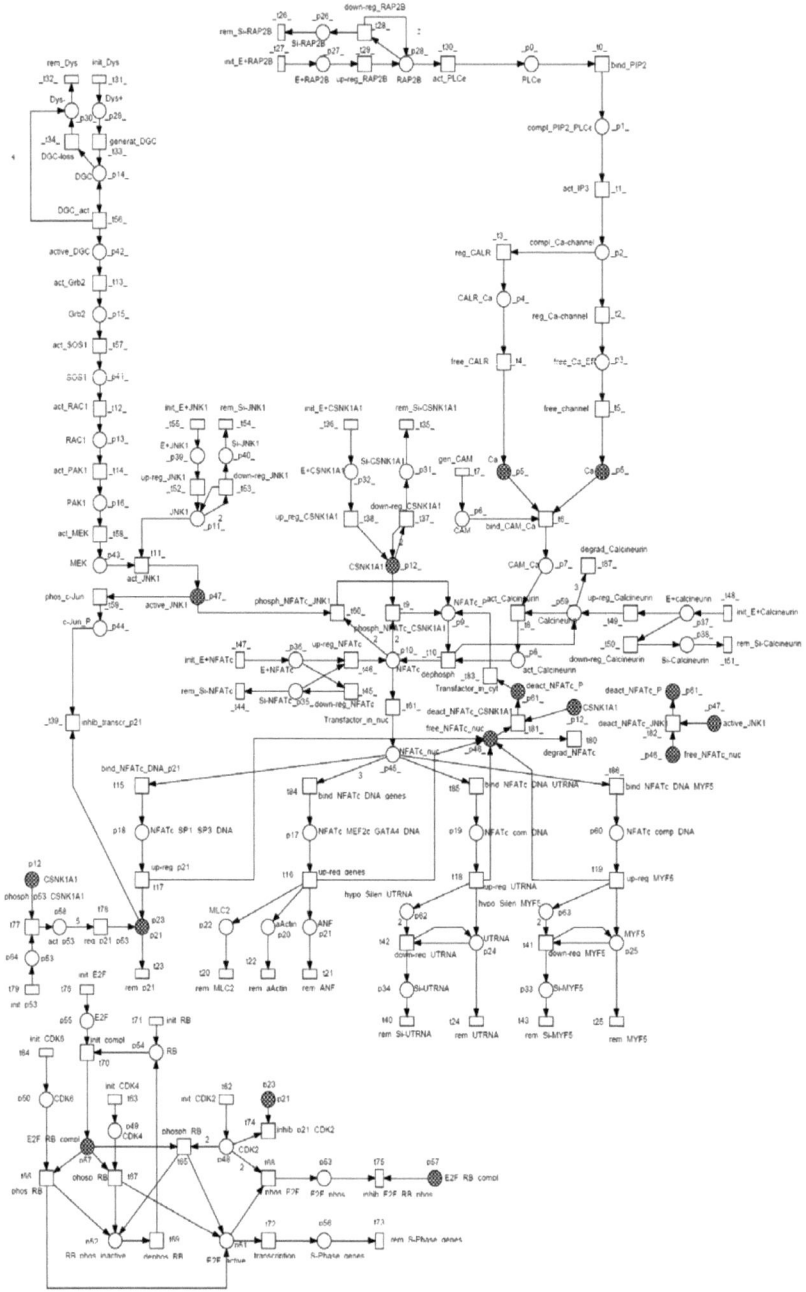

figure 13: The Petri net model of the dystrophin downstream cascade and converging pathways based on the graphically represented signal transduction network in section 4.1.

For modelling of mRNA expression patterns, another structure was introduced for genes whose mRNA expression was experimentally determined. Alterations in gene expression were modelled by 2 places following a suggestion firstly described by Chaouiya et al. (Chaouiya et al., 2004). One place represents up-regulation (enhancing) and the other down-regulation (silencing). Genes that exhibit transitions for up- and down-regulation are listed in table 30.

table 30: Experimentally explored genes and their transitions for up- and down- regulation. The transitions are indicated by a larger arc weight (>1) according to experimental data. For RAP2B, CSNK1A1, JNK1, UTRNA and MYF5 a special subnet structure has been chosen as exemplarily depicted in figure 14.

Gene name	Transition for up-regulation	Transition for down-regulation
RAP2B	t29	t28
CSNK1A1	t38	t37
JNK1	t52	t53
NFATc	t46	t45
Calcineurin	t49	t50
UTRNA	t18	t42
MYF5	t19	t41

Moreover, a place for the gene product was introduced. As shown below, analysis techniques facilitated a differentiation between enhancing and silencing because of this type of modelling. These situations were modelled as special subnets (figure 14) occurring 6 times in the net. There are 3 places representing the different expression states of JNK1: the protein JNK1 (p11.JNK1) by itself, the enhanced JNK1 (p39.E+JNK1), and the silenced JNK1 (p40.Si_JNK1).

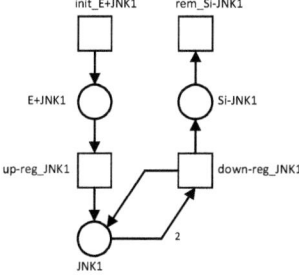

figure 14: A subnet that describes up- and down-regulation of genes. The modelled gene regulation is exemplarily depicted and described for JNK1.

The transition t55.initE+JNK1 produces the protein p11.JNK1 through up-regulation of t52.up_reg_JNK1. The down-regulation of p11.JNK1 took place when at least 2 tokens of p11.JNK1 were produced, indicated by an arc weight of 2. This down-regulation removes tokens from p11.JNK1 through the transition t53.downreg_JNK1 and produces tokens for p11.JNK1 again. However, the structure of the net does not avoid a situation at which both places, p39.E+JNK1 and p40.Si_JNK1, were marked. The regulation of most of the genes (except for several proteins only expressed in developmental stages) including JNK1 does not necessarily mean a complete shut-down or turning-on of its expression. For example, enhancing of JNK1 (p39.E+JNK1) resulted in an increased expression by positive transcription factors comprising a low effect of negative regulators (p40.Si_JNK1) as well as the other way around. This special subnet was also used for RAP2B, CSNK1A1, UTRNA and MYF5 as well as for activation of the DGC (table 29 and table 30). In the case of calcineurin and NFATc, no differences in the mRNA expression levels in patients compared to normal controls were determined. Therefore, an equilibrium between up- and down-regulation of calcineurin and NFATc, respectively, occurred, which was modelled as initiation, up- and down-regulation, and removal of both substances.

4.3.2 Structural analysis of the model

When models grow larger, it becomes difficult to test them for completeness and biological meaning. One way is the simulation of the dynamic network behaviour using ordinary differential equations, and comparing of simulation results with known experimental data. However, this requires the knowledge of all kinetic data. To obtain informations on systems consistency and the general system behaviour, static and dynamic properties of the Petri net were computed and analysed. The following minimal criteria were suggested by [Heiner & Koch (2004)], [Koch & Heiner (2008)]: (1) the net is connected, (2) the net is covered by t-invariants (CTI), and (3) each t-invariant and p-invariant has a biological meaning. Whereas the first criterion can be checked effortlessly, the second and, in particular, the third criterion can become impossible to check in large and complex nets, i.e. dense network graphs. Computation of all t-invariants is feasible, but their analysis is not manually manageable for large amounts of t-invariants. Therefore, MCT-sets and cluster analysis were used for further network decomposition and to explore t-invariants in terms of their biological meaning. The following basic properties were considered and computed using the software INA. The net is connected. It exhibits no places without pre- and/or post-transitions, but does exhibit transitions without pre- or post-places representing the systems interface to the surroundings. The net is unbounded, which means that there are places with possibly infinite token number. The net is live, which indicates that the net is always working continuously and has no dead transitions in the initial marking represented in figure 13.

4.3.2.1 Invariant analysis

t-invariants

The net contained no p-invariant because a gene-regulatory network was considered involving protein-interaction that excluded substance conservation. Substances like ATP, ADP, or AMP were not specially modelled because of their conserved amount and general availability within the cell. A number of 107 t-invariants were computed. The results are given as supplementary material in Grunwald et al. [Supplementary material (Grunwald et al., 2008)]. These t-invariants are graphically represented using different colours in separate files.

The smallest t-invariants consist of 3 transitions, e.g. t-invariant Inv5 includes transitions t48.init_E_Calcineurin, t49.up_reg_Calcineurin, and t87.degrad_Calcineurin. The 4 largest t-invariants, Inv96, Inv97, Inv98, and Inv99, contain 32 transitions each. All of the 4 t-invariants lead to a transcription that belongs to cluster C11. Cluster C11 starts with the DGC downstream pathway which activates up-regulated JNK1, and leads to de-phosphorylation of NFATc by the RAP2B-calcineurin pathway. NFATc is then active, and migrates into the nucleus to mediate transcription of MLC2, aActin, and ANF. In the nucleus, NFATc can be phosphorylated and therefore inactivated by JNK1.

The following transitions are crucial for the network behaviour because of their occurrence in more than 70% of all t-invariants: t0.bind_PIP2, t1.act_IP3, t6.bind_CAM_Ca, t7.gen_CAM, t8.act_Calcineurin, t10.dephosph, t30.act_PLCe, and t61.Transfactor_in_nuc, as well as t83.Transfac_in_cyt which was found in more than 67% of all t-invariants.

Except for one t-invariant (Inv55), all other t-invariants contained both input and output transitions. The t-invariant Inv55 has 4 input transitions and no output-transition. Hence, it describes a cyclic pathway. In this t-invariant, de-phosphorylation of NFATc by calcineurin, which was activated by the RAP2B downstream pathway and phosphorylation by the kinase CSNK1A1, are balanced. Consequently, at no time NFATc was able to migrate into the nucleus and to act as transcription factor. That is an active way of the cell to down-regulate subsequent gene transcription without the need of protein degradation or gene silencing.

RESULTS

MCT-sets

Another analytical approach used are MCT-sets. Transitions in an MCT-set are assumed to be always active together. Thus they should coherently be up- and down-regulated. MCT-sets are interpreted as building blocks leading to a network reduction. The network can be constructed using these building blocks. For example, to perform an exhaustive knockout analysis of the network it suffices to knock out only one arbitrary transition of an MCT-set. A number of 33 MCT-sets were calculated using PInA software (Grafahrend-Belau, 2006). MCT-sets consisting of at least 2 transitions (25) are compiled in table 31. The net with the coloured MCT-sets is given in the supplementary material in Grunwald et al. [Supplementary material (Grunwald *et al.*, 2008)].

table 31: The MCT-sets of t-invariants containing more than one t-invariant. Transitions of one MCT-set were always found together. The MCT-sets M1 and M22 are disconnected MCT-sets.

MCT-set	Composition	Biological interpretation
M1	t0, t1, t6, t7, t8, t10, t30	Initiation of calcium release out of ER
M2	t2, t5	Calcium-channel-mediated calcium release depending on concentration gradient between ER and cytosol
M3	t3, t4	Calcium release regulated by calreticulin
M4	t11, t12, t13, t14, t56, t57, t58	Activation of DGC downstream pathway that activates JNK1
M5	t15, t17	Transcription of p21 by transcription factor complex including NFATc
M6	t16, t20, t21, t22, t84	Transcription of MLC2, aActin, ANF, by NFATc and other factors
M7	t18, t24, t40, t42, t85	Regulated transcription of UTRNA by NFATc and other factors
M8	t19, t25, t41, t43, t86	Regulated transcription of MYF5 by NFATc and other factors
M9	t26, t28	Down-regulation of RAP2B
M10	t27, t29	Up-regulation of RAP2B
M11	t31, t33, t34	Initiation of dystrophin followed by generation of DGC and simulation of DMD by DGC loss
M12	t35, t37	Down-regulation of CSNK1A1
M13	t36, t38	Up-regulation of CSNK1A1
M14	t39, t59	Inhibition of p21 transcription by phosphorylated c-Jun
M15	t45, t47	Down-regulation of NFATc
M16	t49, t87	Up-regulation of calcineurin
M17	t50, t51	Down-regulation of calcineurin
M18	t52, t55	Up-regulation of JNK1
M19	t53, t54	Down-regulation of JNK1
M20	t63, t67	Initiation of CDK4 and phosphorylation of RB by CDK4
M21	t64, t66	Initiation of CDK6 and phosphorylation of RB by CDK6
M22	t68, t71, t75	Initiation of RB, phosphorylation of E2F by CDK2 which inhibits RB phosphorylation (preservation of E2F-RB complex)
M23	t69, t70, t76	Initiation of E2F followed by initiation of E2F-RB complex and de-phosphorylation of RB
M24	t72, t73	Transcription of S-phase gene
M25	t77, t78, t79	Initiation and phosphorylation of p53 by CSNK1A1 which regulates transcription of p21

Except for M1 and M22, all MCT-sets (23) represent connected sub-networks. The MCT-set M1 describes the initiation of the calcium release out of the ER, binding of calcium to calmodulin, activation of calcineurin, and de-phosphorylation of NFATc by calcineurin. Starting with transition t30.act_PLCe followed by transitions t0.bind_PIP2 and t1.act_IP3, the MCT-set stops at place p2.compl_Ca_channel where the pathway splits into 2 possible ways. One way represents the calcium release regulated by calreticulin through the transitions t3.reg_CALR and t4.free_CALR which form MCT-set M2. The other way is the Ca-channel-mediated calcium release, which is dependent on the concentration gradient between ER and cytosol. This pathway is characterised through the transitions t2.red_Ca-channel and t5.free_channel which build MCT-set M3. The MCT-set M1 continued with the transitions t6.bind_CAM_Ca, t7.gen_CAM, t8.act_Calcineurin, and t10.dephosph.

The MCT-set M22 consists of the transitions t68.phos_E2F, t71.init_RB, and t75.inhib_E2F_RB_phos where t71.init_RB is not connected to t68.phos_E2F or t75.inhib_E2F_RB_phos, which are in turn connected. This disconnected MCT-set occurs because of the branching of the pathway represented by the t-invariants Inv1, Inv2, and Inv3. All these t-invariants describe the process of inhibition of the E2F-RB complex via an initiation of RB. They share transitions in that MCT-set, which always occur together but also contain transitions that are exclusively in one of these 3 t-invariants. After generation of RB and E2F and forming of the E2F-RB complex, the pathway can emerge in different ways due to the fact that 3 different kinases were able to phosphorylate the RB protein. Whereas Inv1 describes phosphorylation of RB (t65.phosph_RB) by CDK2, Inv2 includes phosphorylation by CDK4 (t67.phosp_RB), and Inv3 models phosphorylation of RB (t66.phos_RB) by CDK6. The active E2F structure, the phosphorylation of active E2F, and the subsequent pathway have 3 transitions (t68.phos_E2F, t71.init_RB, t75.inhib_E2F_RB_phos) in common. Thus, the kinases CDK2, CDK4, and CDK6, which were responsible for phosphorylation of the same protein, are the reason for the disconnected MCT-set. That means that these transitions work independently.

A number of 15 cases were determined consisting of one transition only (table 32).

table 32: The MCT-sets consisting of only one transition.

MCT-set	Transition	MCT-set	Transition
M26	t80.degrad_NFATc	M34	t32.rem_Dys
M27	t61.Transfactor_in_nuc	M35	t65.phosph_RB
M28	t81.deact_NFATc_CSNK1A1	M36	t46.up-reg_NFATc
M39	t83.Transfac_in_cyt	M37	t62.init_CDK2
M30	t82.deact_NFATc_JNK1	M38	t60.phosph_NFATc_JNK1
M31	t9.phosph_NFATc_CSNK1A1	M39	t23.rem_p21
M32	t48.init E+Calcineurin	M40	t44.rem_Si-NFATc
M33	t74.inhib_p21_CDK2		

Cluster analysis

The application of the established data mining technique, such as the cluster analysis which is a well-known method in phylogenetic tree analysis, is another way of network decomposition and reduction. It provides a better overview of processes taking place and facilitates network validation. The clustering tree calculated is shown in figure 15 (page 97). For clustering of t-invariants, the technique and software as reported in (Grafahrend-Belau, 2006; Grafahrend-Belau et al., 2008a) were applied. The UPGMA algorithm with a threshold of 65% was used for clustering based on the *Tanimoto coefficient* as distance measure between t-invariants. Biological meaning and function of each cluster are described in table 33.

The composition of MCT-sets and transitions of all 34 clusters are given in the supplementary material in Grunwald et al. [Supplementary material (Grunwald et al., 2008)].

table 33: Biological interpretation of t-invariants clusters. Clusters were calculated using UPGMA with a threshold of 65%, based on the *Tanimoto coefficient* as distance measure.

Cluster	Biological interpretation
C1	Initiation and down-regulation of calcineurin
C2	Initiation, up-regulation and degradation of calcineurin
C3	Initiation of dystrophin followed by generation of DCG and simulation of DMD by DGC loss
C4	Initiation, up-/down-regulation of JNK1
C5	Initiation and up-regulation of CSNK1A1, which activates p53, followed by p21 transcription, which inhibits CDK2
C6	Initiation and down-regulation of CSNK1A1, which activates p53, followed by p21 transcription, which inhibits CDK2
C7	DGC downstream pathway, which activates up-regulated JNK1, followed by a c-Jun phosphorylated dependent p21 inhibition, up-regulation of CSNK1A1, which activates p53, followed by p21 transcription
C8	DGC downstream pathway, which activates up-regulated JNK1, followed by a c-Jun phosphorylated dependent p21 inhibition, down-regulation of CSNK1A1, which activates p53, followed by p21 transcription

Cluster	Biological interpretation
C9	DGC downstream pathway, which activates up-regulated JNK1, followed by a c-Jun phosphorylated dependent p21 inhibition, RAP2B-calcineurin pathway, which de-phosphorylates NFATc, followed by a p21 transcription inhibiting CDK2, and phosphorylation of NFATc in nucleus by regulated CSNK1A1 and JNK1
C10	DGC downstream pathway, which activates up-regulated JNK1, followed by a c-Jun phosphorylated dependent p21 inhibition, regulated RAP2B-calcineurin pathway, which de-phosphorylates NFATc, followed by a p21 transcription, and phosphorylation of NFATc in nucleus by JNK1
C11	DGC downstream pathway, which activates up-regulated JNK1, regulated RAP2Bcalcineurin pathway, which de-phosphorylates regulated NFATc, followed by a transcription of MLC2, aActin, ANF, and phosphorylation of NFATc in nucleus by JNK1
C12	DGC downstream pathway, which activates up-regulated JNK1, followed by phosphorylation of NFATc in cytosol by JNK1, regulated RAP2B-calcineurin pathway, which de-phosphorylates regulated NFATc, which does not result in transcriptional activity
C13	DGC downstream pathway, which activates up-regulated JNK1, followed by a c-Jun phosphorylation dependent p21 inhibition, regulated NFATc mediates p21 transcription, followed by a degradation of NFATc in nucleus
C14	Down-regulated RAP2B-calcineurin pathway, which de-phosphorylates NFATc, followed by a p21 transcription inhibiting CDK2, and phosphorylation of NFATc in nucleus by regulated CSNK1A1
C15	Up-regulated RAP2B-calcineurin pathway, including Ca release depending on concentration gradient between ER and cytosol, which de-phosphorylates NFATc, followed by p21 transcription inhibiting CDK2, and phosphorylation of NFATc in nucleus by regulated CSNK1A1
C16	Down-regulated RAP2B-calcineurin pathway, including Ca release depending on calreticulin, which de-phosphorylates NFATc, followed by p21 transcription inhibiting CDK2, and followed by phosphorylation of NFATc in nucleus by regulated CSNK1A1
C17	Up-regulated RAP2B-calcineurin pathway, including Ca release depending on calreticulin, which de-phosphorylates NFATc, followed by a p21 transcription, inhibiting CDK2, and followed by phosphorylation of NFATc in nucleus by regulated CSNK1A1
C18	Regulated RAP2B-calcineurin pathway, which de-phosphorylates regulated NFATc, followed by a transcription of MLC2, aActin, ANF, and phosphorylation of NFATc in nucleus by regulated CSNK1A1
C19	Phosphorylation of NFATc in cytosol by regulated CSNK1A1, regulated RAP2B-calcineurin pathway, including Ca release, depending on concentration gradient between ER and cytosol, which de-phosphorylates regulated NFATc, which does not result in transcriptional activity
C20	Phosphorylation of NFATc in cytosol by regulated CSNK1A1, regulated RAP2B-calcineurin pathway, including Ca release, depending on calreticulin, which de-phosphorylates regulated NFATc which does not result in transcriptional activity
C21	Down-regulated RAP2B-calcineurin pathway, including Ca release depending on concentration gradient between ER and cytosol, which de-phosphorylates NFATc, followed by transcription of UTRNA and MYF5, and phosphorylation of NFATc in nucleus by regulated CSNK1A1
C22	Up-regulated RAP2B-calcineurin pathway, including Ca release, depending on concentration gradient between ER and cytosol, which de-phosphorylates NFATc, followed by transcription of UTRNA and MYF5, and phosphorylation of NFATc in nucleus by regulated CSNK1A1
C23	Down-regulated RAP2B-calcineurin pathway, including Ca release, depending on calreticulin, which de-phosphorylates NFATc, followed by transcription of UTRNA and MYF5, and phosphorylation of NFATc in nucleus by regulated CSNK1A1
C24	Up-regulated RAP2B-calcineurin pathway, including Ca release, depending on calreticulin, which de-phosphorylates NFATc, followed by transcription of UTRNA and MYF5, and phosphorylation of NFATc in nucleus by regulated CSNK1A1
C25	DGC downstream pathway, which activates up-regulated JNK1, regulated RAP2B-calcineurin pathway, which de-phosphorylates regulated NFATc, followed by a transcription of UTRNA and MYF5 and phosphorylation of NFATc in nucleus by JNK1
C26	Initiation and down-regulation of NFATc
C27	Regulated NFATc mediates p21 transcription inhibiting CDK2, followed by degradation of NFATc in nucleus
C28	Regulated NFATc mediates transcription of MLC2, aActin, and AFN followed by degradation of NFATc in nucleus

RESULTS

Cluster	Biological interpretation
C29	Regulated NFATc mediates transcription of MYF5, followed by degradation of NFATc in nucleus
C30	Regulated NFATc mediates transcription of UTRNA, followed by degradation of NFATc in nucleus
C31	CDK6-dependent RB-E2F cell cycle pathway, resulting in transcription of S-phase genes
C32	CDK4-dependent RB-E2F cell cycle pathway, resulting in transcription of S-phase genes
C33	CDK2-dependent RB-E2F cell cycle pathway, resulting in transcription of S-phase genes
C34	RB-E2F cell cycle pathway, inhibited by CDK2-phosphorylated E2F

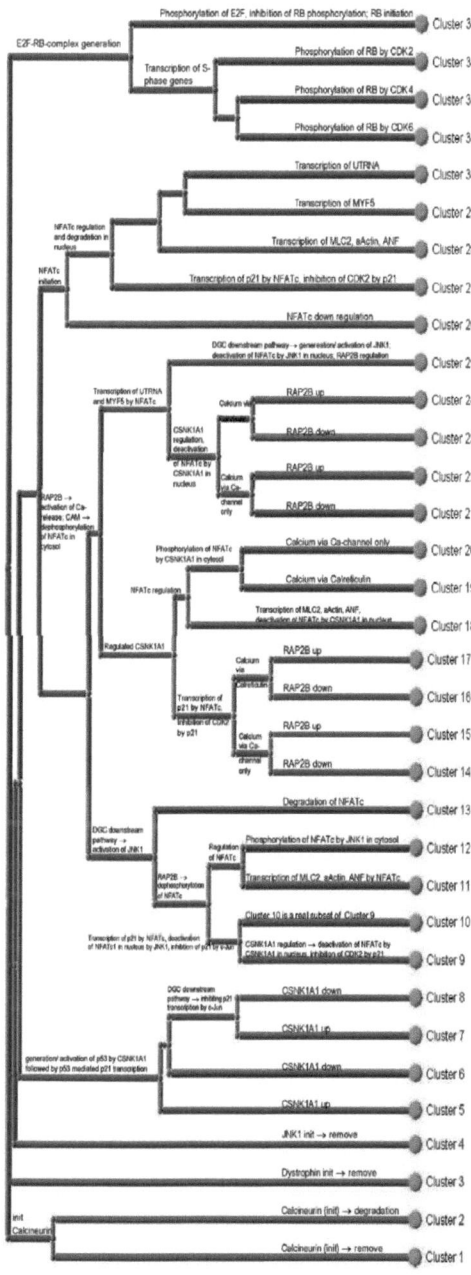

figure 15: The cluster tree. The edges are labelled according to the distinguishing features between clusters before the first branch and the common properties after the first branch.

4.3.2.2 Knockout analysis using Mauritius map

Knockout experiments are a valuable method of gaining insights into molecular processes. The methods used are rather formal and can be applied to any Petri net. On the one hand, findings of the analysis should be consistent with the biological meaning and information on which the model is based. On the other hand, the findings also inspire new questions and experiments for future work.

At first, the knockout impact of a single activity was studied to analyse the knockout behaviour of the model. Activities were represented either by an MCT-set or a single transition. The impact of a knockout of a transition was measured by the percentage of the number of t-invariants affected by its loss. This corresponds to the rate of reduction of the functional diversity or combinatorial complexity of the system.

In table 34, the most important activities are listed as observed in the Petri net model. De-phosphorylation of the transcription factor NFATc (M1), its migration into the nucleus (t61.Transfactor_in_nuc), and transport back to the cytosol (t83.Transfactor_in_cyt) play a crucial role. Whereas the overall impact of a single knockout can be simply studied by deleting the corresponding activity, the dependency of the functional entities on each other can become very complex.

table 34: Most important activities according to their combinatorial knockout impact. The impact of a knockout is measured by the percentage of the number of t-invariants affected by it. Activities with a knockout impact below 20% are not listed.

MCT-set/ transition	Activity	Knockout impact
M1	De-phosphorylation of NFATc	78%
t61	NFATc migrates into nucleus	73%
t83	Transportation of NFATc back to cytosol	67%
t81	Deactivation of NFATc by CSNK1A1	45%
M2	Ca release mediated by Ca-channel	39%
M3	Ca release regulated by calreticulin	39%
M9	Down-regulation of RAP2B	39%
M10	Up-regulation of RAP2B	39%
t32	Removal of dystrophin	37%
t52	Removal of silenced calcineurin	37%
t55	Initiation of enhancer of JNK1	37%
M4	Activation of the DGC pathway	36%
M5	Transcription of p21	36%
M12	Down-regulation of CSNK1A1	29%
M13	Up-regulation of CSNK1A1	29%
M15	Down-regulation of NFATc	29%
t46	Up-regulation of NFATc	28%
t82	Deactivation of NFATc by JNK1	22%

Mauritius maps have been introduced to visualise this relationship (section 1.4.6). The entire Mauritius map for the Petri net model is shown in figure 16. It gives a visual impression of the complexity of functional dependencies in the model. For the sake of clarity, the labels of most of the edges have been dropped. For detailed examination, please refer to the file pn2_0803_fullb.eps in the supplementary material in Grunwald et al. [Supplementary material (Grunwald et al., 2008)]. On inspection of figure 16 it becomes obvious that the interrelations of molecular processes in the gene regulation of DMD are very complex in detail. Some of the edges are drawn as rather thick lines, i.e. they cover large parts of the net, whereas other lines are thin, i.e. they have only local influence.

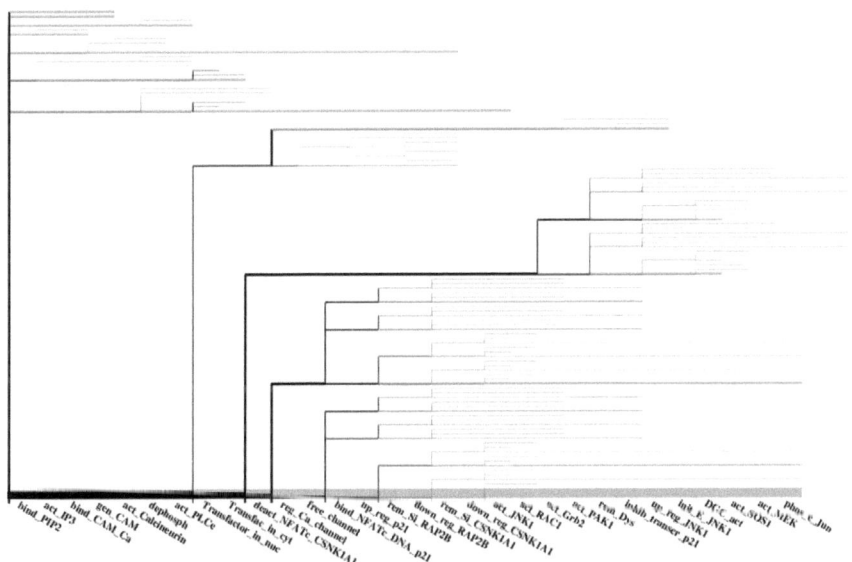

figure 16: The complete Mauritius map of the gene regulation in DMD. It gives a visual impression of the complexity of the functional dependencies in the model. Whereas the interrelations of the molecular processes in the gene regulation are very complex in detail, some of the edges cover large parts of the net (drawn as thick lines). To explore the completely lettered graph, please see the file pn2 0803 fullb.eps in the supplementary material in Grunwald et al. [Supplementary material (Grunwald et al., 2008)].

The dominant part of the Mauritius map is presented in figure 17. All the edges of the tree, which belong to a relative knockout impact below 20%, were dropped. That means, only transitions were considered whose knockout would lead to a destruction of more than 20% of all t-invariants. It was assumed that the importance of a transition or of a set of transitions was related to the percentage of invariant destruction when it is knocked out. Obviously, the right edge of the root represents the most important activity, namely the molecular processes of

RESULTS

M1 covering the transitions t0.bind_PIP2, t1.act_IP3, t6.bind_CAM_Ca, and t7.gen_CAM (table 34). In the model, the central role of M1 corresponds to the key role of the de-phosphorylation of NFATc (t10.dephosph). De-phosphorylation of NFATc relies on the activation of PLCe by RAP2B (t30.act_PLCe), the generation of calmodulin (t7.gen_CAM), the binding of calcium to calmodulin (t6.bind_CAM_Ca), and the activation of calcineurin by calmodulin (t8.act_Calcineurin). These activities are indispensable to de-phosphorylation of NFATc in this model, whereas alternatives exist for other processes of the calcium release from the endoplasmatic reticulum (ER).

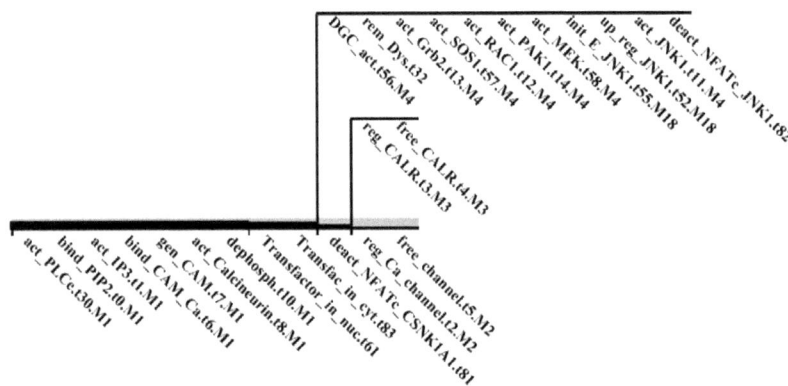

figure 17: Dominant part of the Mauritius map. The edges with relative knockout impact below 20% have been omitted. The dominant part of the Mauritius map describes the interrelation of molecular processes contributing to the de-phosphorylation of transcription factor NFATc. The name of a transition is concatenated with its transition number and MCT-set.

Considering the functional part of the model that depends on de-phosphorylation of NFATc, the most important transitions are t61.Transfactor_in_nuc and t83.Transfactor_in_cyt. These are the labels between the second and the third edge in the Mauritius map (figure 16), and are part of M1. The corresponding molecular processes, more precisely the migration of the transcription factor NFATc into the nucleus (t61.Transfactor_in_nuc) and its transport back to the cytosol (t83.Transfactor_in_cyt), act as an MCT-set. These transitions resulted in identical knockout effects in this context. The left edge denotes the deactivation of NFATc by CSNK1A1 (t81.deact_NFATc_CSNK1A1). De-activation of NFATc is a pre-condition for transportation of NFATc back to the cytosol.

The phosphorylation was mediated either by CSNK1A1 or via the DGC downstream pathway by JNK1. Consequently, molecular processes indispensable for the activation of JNK1 via the DGC downstream pathway are the labels of the right edge. These processes only result in an

MCT-set in the context of phosphorylation of NFATc, but might participate in different ways to other activity groups of the net.

The child node of the activity of CSNK1A1 includes phosphorylation of the transcription factor NFATc (t81.deact_NFATc_CSNK1A1). Since phosphorylated NFATc cannot enter the nucleus, the phosphorylation activity of NFATc would lead to an accumulation of the transcription factor deactivated in the cytosol. The high level of phosphorylated NFATc in the cytosol has to be compensated by an effective de-phosphorylation process, i.e. by an advanced activity of calcineurin. However, the activation of calcineurin is mediated by the calcium regulated calmodulin. One of the main storages of calcium is the endoplasmatic reticulum. Its release into the cytosol can be initiated depending on the calcium concentration gradient between the endoplasmatic reticulum and the cytosol (M2) or by calreticulin dependent processes (M3). Both of these 2 functional groups represent alternative ways to provide free calcium for binding to calmodulin or other proteins. Consequently, the molecular processes of M2 and M3 are the labels of the following right and left edges. A complete knockout of the function of calcineurin only is achievable by a knockout of molecular processes of both MCT-sets.

A special Mauritius map was constructed to study the dependency of a particular function on other transitions rather than the detailed structure of functional dependencies in the entire net. A transition of interest was chosen that was now exhibiting the highest impact by definition. The transition selected was highlighted, and appeared as one label of the left edge of the root. For example, the modified phosphorylation of c-Jun by active JNK1 (t59.phos_c-Jun) is depicted in figure 18. A knockout of this transition caused an impact of around 13% whereas other phosphorylation processes usually exhibited knockout impacts below 5%. In the entire network (figure 16), MCT-set M14 includes phosphorylation of c-Jun (t59.phos_c-Jun) and inhibition of the transcription of p21 (t39.inhib_transcr_p21). Considering the functional subnet only, which was participating in this process, M14 was then a member of a much larger MCT-set as shown on the label of a left edge of the root in figure 18.

RESULTS

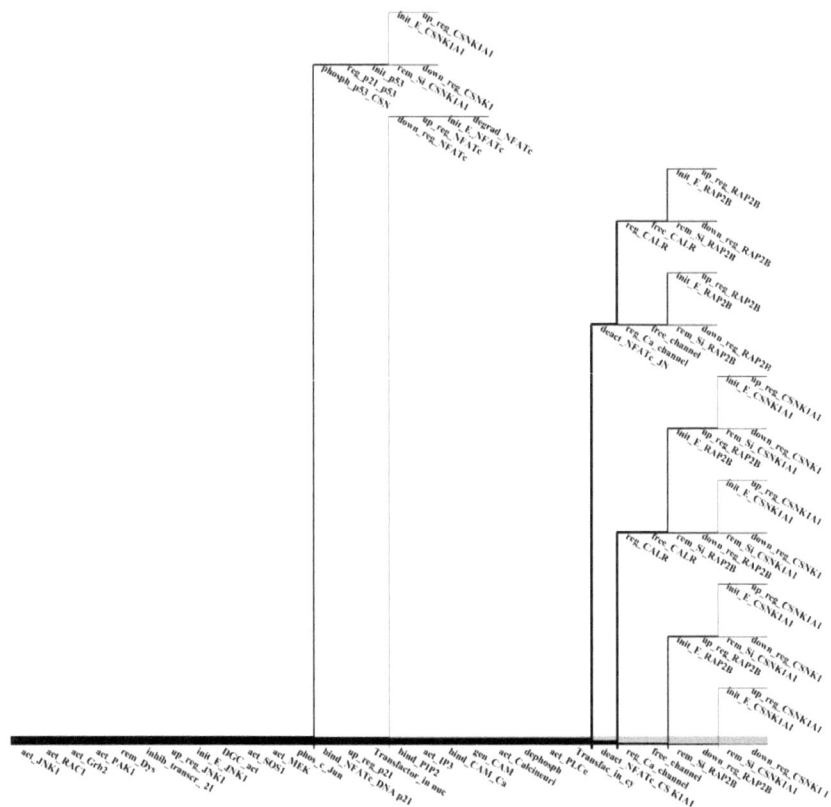

figure 18: Mauritius map for the phosphorylation of c-Jun by active JNK1 (t59.phos_c-Jun). This transition becomes a member of a rather large MCT-set in this sub-network. The interrelations of the molecular processes exhibit a complex structure.

The knockout of any transition of this set would result in a complete knockout of this phosphorylation pathway. The interaction of other knockouts, which partly influence this function only, was exhibiting a rather complex structure. These functional dependencies may be tested by performing systematic knockout experiments.

To summarise, the RAP2B-calcineurin pathway turned out to be one of parts most relevant to network behaviour. This suggested that an influence on calcineurin would also have a vital impact on NFATc. Experiments were proposed using chemical compounds affecting calcineurin and phosphatases in general (section 4.4.4).

4.4 Modulation of gene expression of CSNK1A1 by plasmid vector cDNA as well as siRNA and modification of CSNK1A1 and calcineurin activities using chemical substances

Based on the results of Petri net modelling and subsequent analysis approaches, candidate genes and proteins were selected for manipulation of their expression state and protein activity. In the Petri net model, CSNK1A1 and calcineurin are involved in a signal transduction subnet whose lack had the highest impact in Mauritius map analysis. Since CSNK1A1 was also discovered as significantly differentially expressed at mRNA level in both directions in muscle biopsy tissues, it was intended to experimentally influence its gene regulation with regard to up-regulation and down-regulation. The manipulations of CSNK1A1 and calcineurin were expected to result either in down-regulation of MYF5, p21, and UTRNA via phosphorylation and inactivation of the transcription factor NFATc or vice versa. Although MYF5, p21 and UTRNA were modelled with NFATc as the same transcription factor, it is questionable whether expression regulation of all of them is equivalently governed. For the DMD therapy 2 objectives are of special interest:

Down-regulation of p21 expression to increase myoblast proliferation without a concurrent decrease in UTRNA expression.

Up-regulation of UTRNA expression and maintaining or increasing myoblast proliferation.

Such interventions are feasible using transfection of plasmid vector cDNA constructs or siRNAs and the substitution of chemical compounds. The application of chemical compounds is faster, more convenient and technically simpler than genetic engineering but the influence affects a wide range of proteins rather than a specific one. Accepting this disadvantage, experiments were started using chemical compounds in cell culture.

4.4.1 Characterisation of cell cultures originating from DMD patients and controls

Primary cells derived from DMD patients used in subsequent experiments and patients' clinical features are characterised in table 35. The cell culture 109/03 is equivalent to patient P5 as well as 145/03 to patient P2 (section 3.3.2). The preparation of skeletal muscle biopsies tissue to obtain primary myoblasts cell cultures is described in section 3.3.2.2.

table 35: Summary of clinical features and cultivation specification of DMD-patients and controls used as SkMCs in cell culture. Proliferation data were determined by Stefanie Endesfelder; PhD thesis (Endesfelder, 2004)

Designation		Clinical data			Cell culture	
		Classification	Age at biopsy	Dystrophy	Doubling of population	Subjective impression cultured cells
14/00		normal	2 years	-	4d	normal
18/01		normal	9 years	-	5.5d	normal
43/01		DMD	3 months	no impairment	6d	slightly delayed
77/02		DMD	4 years	progressive, mild course	14d	delayed; myoblasts enlarged
109/03	P5*	DMD	3.5 years	progressive	12d	delayed; myoblasts enlarged
145/03	P2*	DMD	2 years	progressive	7d	slightly delayed

* patients also available for muscle biopsy investigations

4.4.2 Applications governing CSNK1A1 activity and expression

4.4.2.1 Inhibition of CSNK1A1 by IC261 at protein level

IC261 (3-[(2,4,6-Trimethoxyphenyl)methylidenyl]-indolin-2-one) is a specific inhibitor of the protein family casein kinase I. Members of this family include alpha, delta, gamma, and epsilon. The most potent effect of IC261 is observed for casein kinase I delta and epsilon, but it also inhibits CSNK1A1 at higher concentrations (c>10μM). Since CSNK1A1 can phosphorylate and de-activate transcription factor NFATc, a reduction of p21 mRNA levels was assumed. Consequently, an increase in proliferation was expected.

In addition to the determination of cell proliferation, vitality tests were performed using WST 8. All the data were acquired from cells at optimal vitality. IC261 was dissolved in

DMSO. Concentrations are given as final concentrations. For each experiment, untreated as well as DMSO-treated cells were included as negative controls. Unless otherwise noted, proliferation data of cells influenced by chemical compounds were compared to untreated samples. Data were revealed from 3 experiments repeated with n=6 each. The alteration of cell proliferation proposed as consequence of p21 expression depending on IC261 concentration is shown in figure 19. Optimal concentration of IC261 applied to SkMCs turned out to be specific for each cell culture originating from different DMD patients.

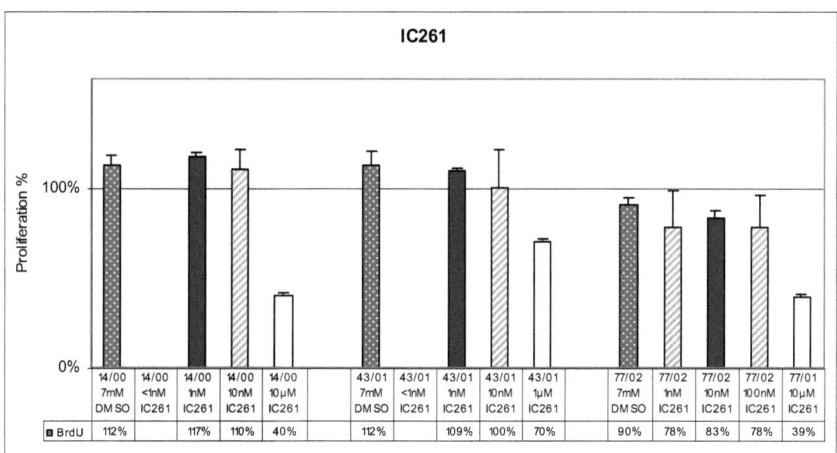

figure 19: Proliferation (BrdU) data of (Dys+) and (Dys-) SkMCs treated with IC261 for 48h. DMSO-treated cells (grey with dots) were compared to untreated cells (100%) to introduce the effect of DMSO to SkMCs. The bar diagram demonstrates developing of proliferation rate depending on IC261 concentration of each treated cell culture. Thus, the dark grey bar indicates best revealed cell proliferation rate whereas striped light grey bars specify concentration next to the optimal one. White bars show the toxic level.
(Dys+) SkMCs: 14/00   (Dys-) SkMCs : 43/01; 77/02

A concentration below 1nM was not performed since no effect of IC261 was expected at that low molar concentration, in particular for CSNK1A1. Both, proliferation and vitality were diminished using concentrations above 1µM, which was the concentration at which an effect was microscopically visible. SkMCs exhibited dramatic but reversible morphological changes. SkMCs usually spread out longitudinally with a smooth cell membrane. Using IC261 (c >1µM) morphology of the cells appeared rounded and jagged. No changes in proliferation were determined for 77/02. A slight but non-significant increase in proliferation was observed for 14/00 and 43/01 using 1nM IC261, but is equivalent to the values of SkMCs treated with DMSO. In the literature, concentrations between 0.1µM and 1µM were suggested as effective (see http://www.emdbiosciences.com/product/d/400090). The results suggest that the effect of IC261 on casein kinase delta and epsilon may be predominant. Casein kinases delta and epsilon are involved in regulating DNA repair and chromosomal segregation. Consequently, IC261 was shown to inhibit cytokinesis by causing a transient mitotic arrest (Behrend et al., 2000). A positive influence of IC261 through inhibition of CSNK1A1 might be not detectable. A specific inhibitor for CSNK1A1 was not available at the time when experiments were performed. For this reason, further experiments were performed using plasmid vector cDNA and siRNA to modulate CSNK1A1 more specifically.

4.4.3 Up-regulation of CSNK1A1 using plasmid vector cDNA

Optimisation of transfection

Most primary cells and in particular SkMCs are very difficult to transfect. Therefore, a transfection technique was needed that provides a high efficiency and low toxicity, especially since (Dys-) SkMCs are more sensitive to transfection reagent and demonstrate a lower proliferation rate than (Dys+) SkMCs. In the literature, only few and partly contradictory data are published regarding transfection of primary human SkMCs (Espinos et al., 2001; Campeau et al., 2001; Pampinella et al., 2002). In this thesis, numerous transfection techniques were examined, such as Fugene HD® and Fugene 6® (Roche), Effectene™ (Qiagen), Lipofectamine 2000® (Invitrogen) and Targefect® (Targeting Systems) as well as the electroporation system amaxa (amaxa). Amaxa is an electroporation system using cell type specific solutions and is also termed nucleofection. Quenneville et al. published a successful introduction of DNA plasmids containing an enhanced GFP (eGFP) construct by nucleofection (Quenneville et al., 2004). Fugene HD®, Fugene 6® and Effectene™ are non-liposomal lipid formulations that are used in transfection approaches of a broad range of cell types and cause minimal cytotoxicity.

Two different cDNA plasmid vectors were applied for transfection. Both of them carry a cDNA insert encoding for CSNK1A1. The cDNA plasmid purchased from Genecopoeia is 7kb in size, also contains a neomycin cassette and GFP as reporter gene connected via an IRES site to the CSNK1A1 insert. However, a poly-A signal is lacking (http://www.genecopoeia.com). The second CSNK1A1 vector was obtained from Origene carrying a full length cDNA including a poly-A signal but not a reporter gene. Controls included positive controls using eGFP plasmid vector at 2kb in size (eGFP; supplied with the amaxa kit) and negative controls treated with the transfection reagent or system without cDNA plasmids.

Transfection experiments using Lipofectamine 2000® and Targefect® did not succeed since no fluorescence signal and negative Real-Time PCR results were received. Both reagents were not considered in the further report.

Fluorescence data were only received from transfections using Genecopoeia cDNA plasmid vector and Effectene^{TM}, Fugene 6® and the amaxa system. Transfection results are exemplarily shown for Effectene^{TM} used in the (Dys+) SkMCs 14/00 (figure 20 - 22). Equivalent results were revealed for transfections using Fugene 6® and the amaxa system. The following experiments were performed:

(1) Negative controls: Cells treated with Effectene^{TM}, but without DNA, demonstrated a faint auto-fluorescence when exposed to FITC filter (figure 20b, but also showed less toxicity of the reagent (figure 20a).

(2) Positive controls: SkMCs were transfected using the 2µg amaxa control vector which resulted in bright fluorescent cells, but with low efficiency (figure 21b). The toxicity was as low as shown for the negative control (figure 21a).

(3) CSNK1A1 transfection: Fluorescence of transfected cells appeared very faint similar to untransfected cells (figure 22b). Cytotoxicity was increased compared to the negative control. That may be due to an increased amount of plasmid DNA used in these experiments. The number of cells decreased during transfection and cells exhibited more inclusions (figure 22a).

RESULTS

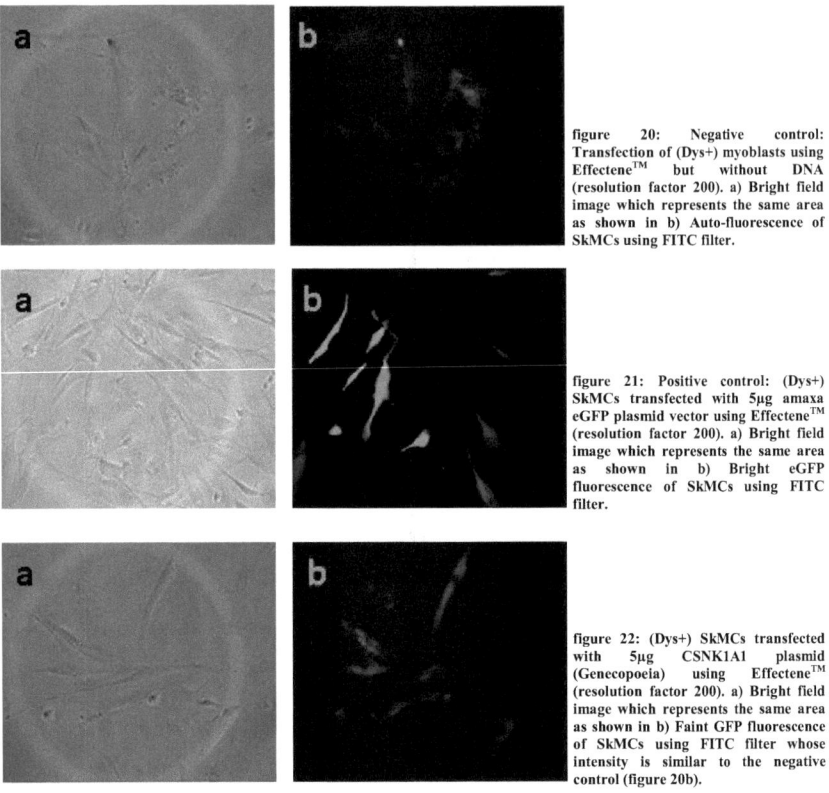

figure 20: Negative control: Transfection of (Dys+) myoblasts using Effectene™ but without DNA (resolution factor 200). a) Bright field image which represents the same area as shown in b) Auto-fluorescence of SkMCs using FITC filter.

figure 21: Positive control: (Dys+) SkMCs transfected with 5µg amaxa eGFP plasmid vector using Effectene™ (resolution factor 200). a) Bright field image which represents the same area as shown in b) Bright eGFP fluorescence of SkMCs using FITC filter.

figure 22: (Dys+) SkMCs transfected with 5µg CSNK1A1 plasmid (Genecopoeia) using Effectene™ (resolution factor 200). a) Bright field image which represents the same area as shown in b) Faint GFP fluorescence of SkMCs using FITC filter whose intensity is similar to the negative control (figure 20b).

To summarise, for Effectene™, Fugene 6® and the amaxa system, a sufficient GFP signal was not detected.

According to published data, the results suggested that too low or ineffective transfection rates were responsible for the negative outcome, and a detection limit of wild type GFP, in particular if expressed via an IRES site. In the literature, experiments that investigated the visibility of eGFP fluorescence depending on the received copy number found that injections of 10^3 cDNA plasmid copies per cell only resulted in a faint eGFP fluorescence signal in 30-50% of the cells. For a bright signal, a plasmid DNA copy number of 10^5 per cell was required (Schindelhauer and Laner, 2002). The Genecopoeia CSNK1A1 cDNA plasmid construct carries a wild type GFP, which is less fluorescent than eGFP and is expressed via an IRES site. So it has been proposed that the transfected copy number is too low for a detectable

GFP signal, but might be sufficient for up-regulation of CSNK1A1 expression. For the determination of the transfection efficiency, quantitative Real-Time PCR was applied to examine not only the absolute copy number of transfected cDNA plasmid vectors but also relative up-regulation of CSNK1A1. At DNA level, 2 primer pairs were used. The first primer pair detected the cDNA insert of the vector as well as 2 known endogenous pseudo-genes of the insert as found at http://www.pseudogene.org. Pseudogenes of CSNK1A1 would explain a positive transfection result of untransfected controls. Since the cells used in all the experiments originated from the same donor, the whole genome and consequently the number of pseudogenes were identical and disregarded. The second pair led to amplification of genomic DNA, and was applied to correct variations of the amount of endogenous DNA utilised in each reaction. A standard curve was calculated with an efficiency of 1.99 for both cDNA plasmid vectors used and is exemplarily shown for the vector purchased from Genecopoeia (figure 23). The adjusted results of all samples were applied to the standard curve to retrieve the copy numbers of each experiment.

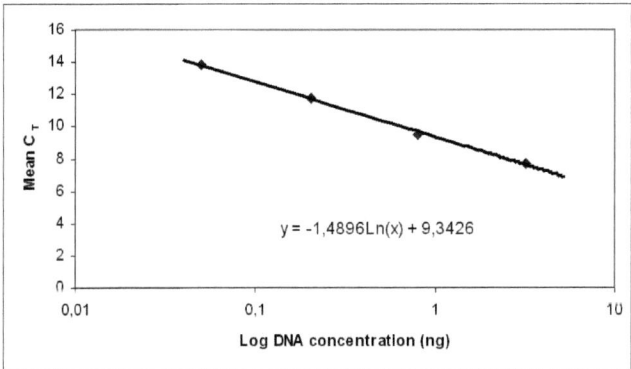

figure 23: Standard curve of cDNA plasmid vector CSNK1A1 purchased from Genecopoeia with an efficiency of 1.99. The equation was used to calculate the copy numbers of transfected plasmid DNA.

The results of the comparison of transfection efficiencies for Fugene HD®, Fugene 6®, Effectene^{TM} and the amaxa system are presented in figure 24, which gives the copy numbers of each sample. For Fugene HD®, different ratios of transfection reagent : DNA are exemplarily given, and were omitted for the other transfection methods. The optimal transfection efficiency was obtained with Fugene HD® : DNA ratio of 6:2 and 200% complex volume at which a total copy number of $2,4 \times 10^{10}$ of cDNA plasmid vectors was determined, isolated from 4×10^5 primary SkMCs (6×10^4 copies/cell). Although, this value appears extremely high at first, such a copy number was determined by Schindelhauer in 2002 for several primary cells (Schindelhauer and Laner, 2002). Thus, the values as shown in figure 24 can be interpreted as realistic. By contrast, the negative controls, treated with Fugene HD® only, still displayed a weak signal in Real-Time-PCR which is thought to be resulting from 2 known pseudogenes of the insert rather than from cDNA plasmid vectors.

Finally, a decrease of volume of Fugene HD® or a complex volume of 400% reduced the transfection efficiency. The latter may also be a result of toxicity. By contrast, an increase of Fugene HD® and DNA (12:4) also revealed satisfying transfection rates but only for a complex volume of 100%.

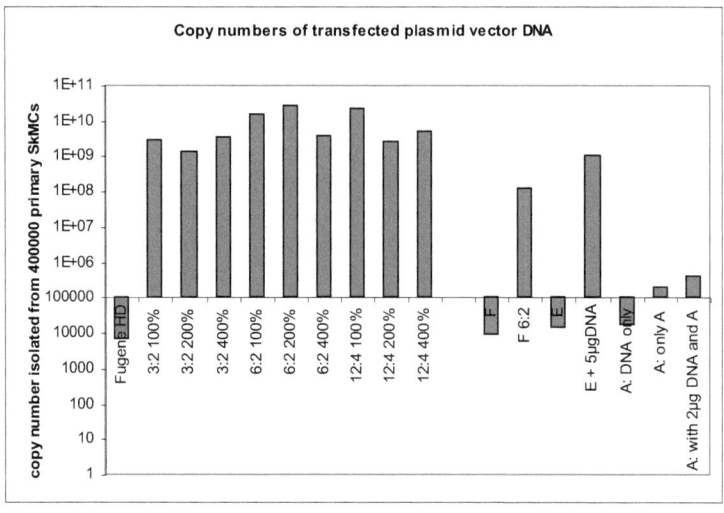

figure 24: Plasmid DNA copy numbers isolated from 4 x 105 primary SkMCs derived with transfection reagents Fugene HD®, Fugene 6 ® (F) and Effectene™ (E) as well as with amaxa (A).

The reagents Effectene™, Fugene 6® and the amaxa system reached a very low percentage of detected vector copies compared to Fugene HD®. Even though the transfection reagent Effectene™ showed the best efficiency of these 3, the transfected plasmid copy number was 1/25 of the highest Fugene HD® transfection result. But in addition it revealed a considerable toxicity in contrast to Fugene HD®. In general, the impression of toxicity was microscopically monitored 24h after incubation of cells with the Fugene HD®-DNA-complex as well as the other reagents and systems. No toxicity could be determined for the Fugene HD®-DNA-complex volumes of 100% and 200% and less toxicity for 400%, whereas 30-40% of the cells died during transfection using the reagents Effectene™ and Fugene 6® as well as the amaxa system (data not shown).

Transfection results using CSNK1A1 plasmid cDNA vector from Genecopoeia
Having the transfection efficiency rated as being satisfactory, optimal transfection conditions were performed to study the expression rate of CSNK1A1, the influence of CSNK1A1 up-regulation on proliferation and vitality of the transfected cells as well as on the RNA expression level of MYF5, p21, and UTRNA. The results are summarised in table 36. Expression and proliferation results were compared to Fugene HD®-treated cells (100%). An optimal transfection was determined using the Fugene HD® : DNA ratio of 6:2 with 200% complex volume which resulted in different amounts of entered plasmid DNA (in ng) and a 1,5-16fold increase of expression levels of CSNK1A1 in SkMCs. The DNA uptake and the resulting up-regulation of CSNK1A1 appeared to be cell type specific. No correlation between inserted plasmid DNA and up-regulation of CSNK1A1 was observed, in particular with regard to the number of patients and controls. The (Dys+) cells 14/00 showed the most elevated expression level of CSNK1A1 whereas the lowest up-regulation of CSNK1A1 was investigated for SkMCs 77/02. Interestingly, these cell cultures showed the lowest and highest, respectively, population doubling time (table 35), which might be the cause for the variation in DNA uptake. The proliferation of (Dys+) cells was not altered, but was slightly reduced in (Dys-) cells except for cell culture 145/03. This cell culture demonstrated a decrease to 34% compared to Fugene HD-treated cells. But changes in proliferation were obviously not associated with a reduction of p21 expression. Since CSNK1A1 is able to phosphorylate and inactivate the transcription factor NFATc, the mRNA expression levels of NFATc target genes, MYF5, p21, UTRNA, were determined by Real-Time-PCR (table 36), and were expected to be reduced. However, a correlation between an up-regulation of CSNK1A1 and mRNA expression levels of any of the genes could not be discovered. The calculation of the relative expression levels using transfected and Fugene HD-treated SkMCs resulted in equal values compared to relative quantification of tranfected and untreated SkMCs.

RESULTS

table 36: Transfection experiments in (Dys+) and (Dys-) cells using cDNA plasmid vector from Genecopoeia. After 48h of transfection, the optimal ratio and the appropriate transfection efficiency were revealed in cell cultures of DMD patients and controls. The proliferation rate was determined using the BrdU test and compared to Fugene HD®-treated cells (100%). The mRNA expression levels of CSNK1A1 and the NFATc target genes MYF5, p21, and UTRNA in transfected cells were compared to Fugene HD-treated cells (corresponds to 1 and 100%, respectively). A comparison with untreated SkMCs resulted in the same outcome.

		Analysis on proliferation and mRNA expression level	Cell type					
			(Dys+)		(Dys-)			
			14/00	18/01	43/01	77/02	109/03	145/03
Genecopoeia cDNA Plasmid vector CSNK1A1		Optimal ratio (Fugene HD®:DNA)	6:2 200%	6:2 200%	6:2 200%	6:2 200%	6:2 200%	6:2 200%
		Transfection efficiency (ng)	15ng	9ng	0.3ng	0,8ng	1,6ng	12ng
		Up-regulation of CSNK1A1 expression (Fugene HD®-treated cells = 1)	16x	2,5x	6x	1,5x	2,8x	2,8x
		Proliferation (Fugene HD®-treated cells = 100%)	94%	100%	87%	73%	65%	34%
	Real-Time-PCR	MYF5	936%	176%	143%,	100%	48%	467%
		p21	472%	105%	144%	75%	87%	487%
		UTRNA	267%	97%	51%	360%	44%	134%

An up-regulation of CSNK1A1 was demonstrated, but did not suffice when the cDNA plasmid vector purchased from Genecopoeia was used. This may be due to an incomplete cDNA insert which lacks a Poly-A signal. For this reason, another cDNA plasmid vector carrying a cDNA insert for CSNK1A1 from Origene was applied in subsequent experiments.

Transfection results using CSNK1A1 plasmid cDNA vector from Origene

The transfection results using the Origene cDNA plasmid vector are summarised in table 37. The expression and proliferation results were compared with Fugene HD-treated cells (100%). An optimal transfection was determined using a Fugene HD® : DNA ratio of 6:2 with 200% complex volume which resulted in different amounts of entered plasmid DNA (in ng) and a 1,5-82fold increase of expression levels of CSNK1A1 in SkMCs. In comparison to Genecopoeia results, the transfection efficiencies and the up-regulation of CSNK1A1 expression was considerably improved using the Origene cDNA plasmid vector. The transfection efficiencies were comparable for all used SkMCs. As shown for Genecopoeia plasmid DNA, no clear correlation between transfection efficiency and up-regulation of CSNK1A1 was observed. Again, the (Dys+) cells 14/00 showed the most elevated expression level of CSNK1A1 whereas 77/02 exhibited the lowest up-regulation of CSNK1A1. For both (Dys+) and (Dys-) cells, a correlation was not determined for transfection efficiency and proliferation rate. Contrary to expectation, the proliferation was not elevated but slightly reduced, and that was not a consequence of p21 up-regulation. Overall, a correlation between an up-regulation of CSNK1A1 and mRNA expression levels of MYF5, p21, and UTRNA was not detected. Equal values were determined by the calculation of the relative expression of transfected and Fugene HD®-treated SkMCs compared to transfected and untreated SkMCs.

table 37: Transfection experiments in (Dys+) and (Dys-) cells using cDNA plasmid vector from Origene. After 48h of transfection, the optimal ratio and the appropriate transfection efficiency were revealed in cell cultures of DMD patients and controls. The proliferation rate was determined using the BrdU test and compared with Fugene HD®-treated cells (100%). The mRNA expression levels of CSNK1A1 and the NFATc target genes MYF5, p21, and UTRNA in transfected cells were compared with Fugene HD®-treated cells (corresponds to 1 and 100%, respectively). A comparison with untreated SkMCs resulted in the same outcome.

	Analysis on proliferation and mRNA expression level		Cell type					
			(Dys+)		(Dys-)			
			14/00	18/01	43/01	77/02	109/03	145/03
Origene cDNA Plasmid vector CSNK1A1	Optimal ratio (Fugene HD®:DNA)		6:2 200%	6:2 200%	6:2 200%	6:2 200%	6:2 200%	6:2 200%
	Transfection efficiency (ng)		14ng	14ng	8ng	14ng	16ng	22ng
	Up-regulation of CSNK1A1expression (Fugene HD®-treated cells = 1)		82x	7x	8,5x	1,5x	4,7x	5,5x
	Proliferation (Fugene HD®-treated cells = 100%)		78%	61%	50%	82%	66%	45%
	Real-Time-PCR	MYF5	1200%	200%	500%	84%	41%	272%
		p21	662%	116%	106%	107%	80%	192%
		UTRNA	479%	111%	200%	42%	35%	131%

In summary, the transfections succeeded in both experiments using 2 different cDNA plasmid vectors. However, substantial variations between different DMD patients, and also between both controls, were remarkable. But neither the expected down-regulation of MYF5, p21, and UTRNA nor a general increase in proliferation due to reduced p21 expression levels was detected through a CSNK1A1 up-regulation.

The CSNK1A1 expression of the DMD patient with mild phenotype demonstrated an increased mRNA level compared to his elder brother, but reduced CSNK1A1 expression compared to 5 other typical DMD patients (see section 4.2.2).

Therefore, experiments considering an up-regulation of CSNK1A1 were complemented with the study of CSNK1A1 down-regulation using siRNA.

4.4.3.1 Down-regulation of CSNK1A1 using siRNA

The experiments were similarly conducted as described for the up-regulation of CSNK1A1. The results are summarised in table 38. Expression and proliferation results were compared to X-tremeGene® siRNA transfection reagent treated cells (100%) and in case of proliferation also to cells added to pure siRNA (100%). The optimal transfection was determined for an

X-tremeGene® : siRNA ratio of 10:1 with 100% complex volume. The transfection resulted in differently decreased expression levels (0,08-1,04fold) of CSNK1A1 in SkMCs. The CSNK1A1 mRNA expression levels of both (Dys+) cell cultures were not affected by a siRNA transfection. By contrast, the down-regulation of CSNK1A1 in (Dys-) cells succeeded and was at the lowest after 48h of transfection. Since CSNK1A1 is able to de-activate NFATc by phosphorylation, an increase in transcription of MYF5, p21, UTRNA was expected. This correlation was not determined in all cell cultures. Furthermore, the proposed increase in p21 expression was expected to lead to a reduced proliferation rate. Except for one control cell culture, the cell proliferation of SkMCs differed in both directions depending on the sample used for comparison. If transfected SkMCs were compared with SkMCs treated with transfection reagent alone, a correlation between proliferation and down-regulation of CSNK1A1 expression was not discovered. If these cells were compared with SkMCs treated with pure siRNA, an elevated proliferation rate was detected. Moreover, this discrepancy disappeared after 72h of transfection. However, a down-regulation of CSNK1A1 was not detected for cells treated with siRNA alone. These data suggest that the transfection reagent might influence the vitality of the cell without visible changes in cell morphology resulting in an inhibition of cell proliferation. However, an immune response cannot be excluded. Calculated values of the relative expression of transfected and siPur-treated SkMCs did not differ to the results determined by relative quantification of untreated and X-tremeGene® transfection reagent treated SkMCs.

table 38: Transfection experiments in (Dys+) and (Dys-) cells using siRNA against CSNK1A1. After 48h of transfection, the optimal ratio and the down-regulation of CSNK1A1 were revealed in cell cultures of DMD patients and controls. The proliferation rate was determined using the BrdU test. The results were compared with X-tremeGene®-treated (100%) as well as with siPur-treated cells (100%). The mRNA expression levels of CSNK1A1 and the NFATc target genes MYF5, p21, and UTRNA in transfected cells were compared with siPur-treated cells (corresponds to 1 and 100%, respectively).

			Cell type					
	Analysis on proliferation and mRNA expression level		Dys+		Dys-			
			14/00	18/01	43/01	77/02	109/03	145/03
siRNA transfection affecting CSNK1A1	Optimal ratio (X-tremeGene® : siRNA)		10:1	10:1	10:1	10:1	10:1	10:1
	Down-regulation of CSNK1A1 expression (siPur-treated cells = 1)		1,04x	0,88x	0,08x	0,12x	0,45x	0,25x
	Proliferation	siPur compared to untreated (100%)	130%	103%	127%	139%	130%	120%
		compared to X-tremeGene®-treated (100%)	78%	97%	111%	67%	68%	75%
	Real-Time-PCR	MYF5	300%	125%	130%	72%	120%	96%
		p21	131%	84%	100%	87%	84%	126%
		UTRNA	140%	68%	200%	83%	56%	78%

In summary, a down-regulation of CSNK1A1 was successfully obtained for DMD patients, but not for both controls. The results noticably varied between each DMD patient and controls. The expected relation between a CSNK1A1 down-regulation, a diminished proliferation and increased expression levels of MYF5, p21, and UTRNA could not be demonstrated.

That missing relation was tested for the specific up- and down-regulation of CSNK1A1 as well as the unspecific regulation using IC261. It was suspected that the modulation of CSNK1A1 leads to alterations of and/or interventions in certain other signalling pathways rather than a specific influence of NFATc. In the literature, several other proteins were suggested as substrates for CSNK1A1, e.g. p53 (Knippschild et al., 2005a; Knippschild et al., 2005b).

This means that neither up-regulation nor down-regulation of CSNK1A1 resulted in an alteration of NFATc target gene expression, i.e. MYF5, p21, and UTRNA. Moreover, proliferation and UTRNA expression were not positively influenced, which was the aim of these approaches as modelled in the NFATc de-activation sub-net of the signalling network.

NFATc is modelled and found to be primarily de-phosphorylated and activated by calcineurin (Chakkalakal et al., 2003). The analysis of the Petri Net by means of Mauritius maps showed that a theoretical knockout in the RAP2B-calcineurin sub-net leads to a malfunction of 78% of the net. Consequently, this sub-net has the highest impact on the net's behaviour. Therefore, applications modulating the calcineurin activity by chemical substances were performed.

4.4.4 Modulation of calcineurin signalling

The calcineurin signal transduction pathway has repeatedly been shown to be essential for skeletal muscle growth (Michel et al., 2004; Stupka et al., 2004; Chakkalakal et al., 2006; Michel et al., 2007; Stupka et al., 2008). Among others, the activity of the phosphatase calcineurin can be influenced by several chemical substances, e.g. okadaic acid, CsA, and deflazacort.

Subsequent experiments were designed as described for IC261 (see section 4.4.2.1). Briefly, optimal concentrations of each drug were identified for each patient and control using proliferation and vitality tests. The experiments were performed at optimal vitality of the SkMCs. DMSO was the solvent for all drugs. Concentrations are given as final concentrations. Negative controls include untreated as well as DMSO-treated cells. If not otherwise indicated, proliferation data of cells influenced by chemical compounds were compared to untreated samples. The data were revealed from 3 repeated experiments with n=6 each. Because of the restricted number of patients and controls, a value of more than 20% was considered as an increase in cell proliferation.

The alterations of cell proliferation depending on the concentration of the chemical substances are exemplarily shown for okadaic acid. Similar experiments were also performed for CsA and deflazacort but the range of molar concentration varied. The depiction of these data was left out to concentrate on more relevant data. In either case, optimal concentration of each substance applied to SkMCs turned out to be specific for each cell culture originating from different DMD patients.

4.4.4.1 Okadaic acid

The toxin okadaic acid is a potent and selective inhibitor of serine/threonine-specific protein phosphatases, in particular protein phosphatases 1 (PP1) and 2A and to a lesser extent calcineurin, and is used in studies of de- and phosphorylation processes (Bialojan and Takai, 1988; Dounay and Forsyth, 2002). Calcineurin de-phosphorylates and activates NFATc, which would lead to an enhanced transcription of its target genes.

Since treatment with okadaic acid was assumed to inhibit calcineurin dependent NFATc activation, a decrease in mRNA expression of NFATc target genes MYF5, p21, and UTRNA was expected. An increase in proliferation was suggested as a consequence of p21 reduction. A concurrent decrease of UTRNA expression was not assumed as shown once (Rodova et al., 2004).

The figure 25 shows cell proliferation rates determined after 48h of incubation. Alterations of cell proliferation have been assumed to be a consequence of changes in p21 expression dependent on okadaic acid concentrations.

A concentration below 0.1nM was not performed since no effect of okadaic acid was expected at that low molar concentration. In the literature, effective concentrations were specified above 0.1nM (http://www.merckbiosciences.co.uk). Both, proliferation and vitality were dramatically diminished at concentrations exceeding 10nM and for the SkMCs 18/01 already at concentration of 3nM of okadaic acid. Additionally, (Dys-) cells were more sensitive to the drug so that lower optimal concentrations were determined.

RESULTS

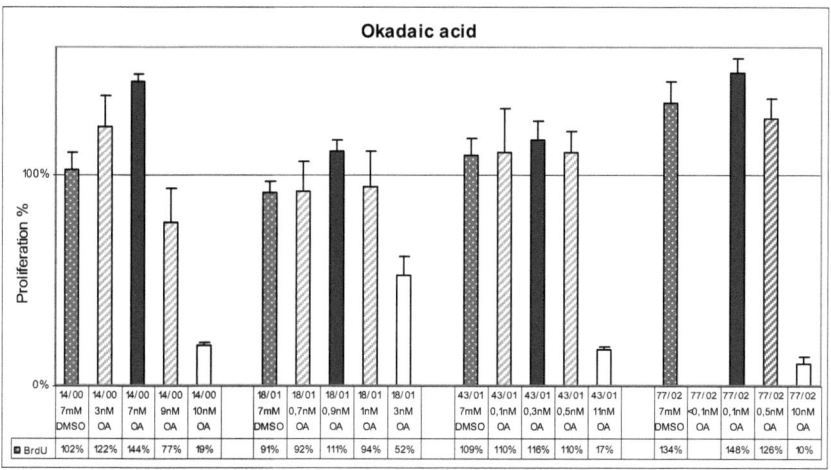

figure 25: Proliferation (BrdU) data of (Dys+) and (Dys-) SkMCs treated with okadaic acid for 48h. DMSO-treated cells (grey and dotted bars) were compared with untreated cells (100%) to introduce the effect of DMSO to SkMCs. The bar diagram demonstrates the developing proliferation rate depending on okadaic acid concentration of each treated cell culture. The black bars indicate the best cell proliferation rate gained whereas shaded bars specify concentrations next to the optimal one. The white bars show the toxic level.
[OA = okadaic acid]   (Dys+) SkMCs: 14/00; 18/01   (Dys-) SkMCs : 43/01; 77/02

The figure displays values of DMSO-treated cells, which exceed proliferation rates of untreated SkMCs (=100%) in some cases. In the following, to exclude the effect of DMSO on SkMCs, the changes in proliferation are specified as the difference between DMSO and okadaic acid treated cells. The (Dys+) myoblasts 14/00 showed a higher increase in proliferation than (Dys-) SkMCs. In comparison to DMSO-treated cells, the elevation in proliferation was determined at 42% for 14/00 and at 20% for 18/01. By contrast, the (Dys-) SkMCs 43/01 revealed no induction (7%) and 77/02 showed a slight stimulation of 14% in proliferation. This positive impact on proliferation disappeared after another 24h of incubation. It seems reasonable to assume that the substance accumulates and reaches its toxic concentration. Obviously, the difference in concentration that results either in increased cell proliferation or toxic events is very minimal. Taking into account this fact, an application of okadaic acid to DMD patients is infeasible even if the proliferation of myoblasts was elevated. In summary, if an optimal concentration was found for the particular SkMCs, an increase in proliferation was demonstrated as expected and modelled in the Petri net. But for all that, the toxicity of okadaic acid is high and very close to its beneficial concentration. For this reason, another chemical substance, CsA, was introduced for inhibiting calcineurin in further experiments.

4.4.4.2 Cyclosporin A

In addition to the anti-inflammatory effects, it is well known that the calcineurin inhibitor CsA is involved in the NFATc signal transduction pathway (St-Pierre et al., 2004; Stupka et al., 2004). CsA was reported to benefit muscular dystrophy once (Sharma et al., 1993). But, contrary results using mdx mice were also reported (Stupka et al., 2004). Currently, a clinical study using CsA in DMD patients is being conducted at the paediatric clinic in Freiburg, Germany (Korinthenberg, 2008).

As described for okadaic acid (section 4.4.4.1), an increase in proliferation with a simultaneous decrease in NFATc-mediated expression of p21 and MYF5 was expected. A reduction of UTRNA mRNA level by calcineurin inhibition is also supposed and was shown in C57BL/6 mice by Chakkalakal et al. (Chakkalakal et al., 2003) and would be an undesired effect.

The SkMCs were treated using a wide range of concentrations of the drug as well as at different time points to identify the optimal dilution in terms of an increase in proliferation and vitality (table 39). The range of optimal concentration of each patient and control individual slightly diverges. A toxic concentration was reached above 1 µM.

table 39: Optimal concentrations of CsA after 48h treatment investigated in SkMCs using proliferation and vitality tests as well as expression studies.

Cells		CsA
Dys+	14/00	5nM
	18/01	10nM
Dys-	43/01	10nM
	77/02	50nM
	109/03	1nM
	145/03	50nM

The cell proliferation results of (Dys+) and (Dys-) SkMCs treated with CsA are shown in figure 26a. Except for 43/01, the cell proliferation was increased for all (Dys-) cell cultures (>20%). A slight elevation was observed for 43/01 and the (Dys+) SkMCs 18/01 (>10%). For the BrdU cell proliferation assay, a variance of 10% was determined for the absorbance values by the manufacturer. The second control myoblasts cell culture 14/00 remained unchanged. The positive impact on cell proliferation was diminished after another 24h incubation as described for okadaic acid (section 4.4.4.1).

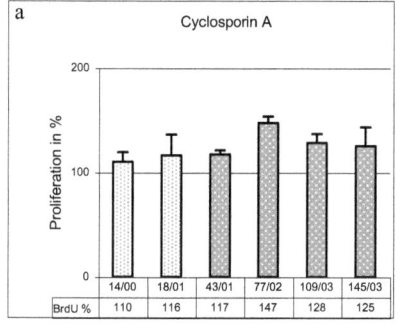 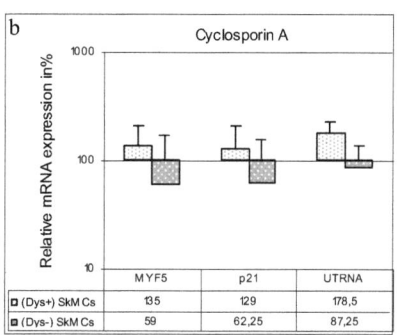

figure 26: (Dys+) and (Dys-) SkMCs treated with CsA for 48h.
a) The altered proliferation rates were normalised to DMSO-affected SkMCs. The cells applied to the solvent DMSO are indicated by the 100% curve. b) The mRNA expression of MYF5, p21 and UTRNA were examined in myoblasts treated with the optimal concentration of CsA. The results were calculated relatively to DMSO-affected SkMCs (=100%).
(Dys+) SkMCs: 14/00; 18/01   (Dys-) SkMCs : 43/01; 77/02; 109/03; 145/03

The (Dys+) and (Dys-) SkMCs treated with the optimal concentrations of CsA were studied at their mRNA levels of MYF5, p21 and UTRNA (figure 26b). The myoblasts affected by DMSO are indicated by the 100% curve. The quantification of CsA-treated cells was relatively calculated to DMSO-affected SkMCs as well as untreated SkMCs (data not shown), and produced equal results. Because of the restricted number of patients and controls, a relative difference of more than 200% and less than 50% was considered as an increase and decrease, respectively, in mRNA expression. Contrary to expectation, for all 3 genes, no alterations in mRNA expression were determined. Consequently, the increase in proliferation was not a consequence of a reduction of p21 expression.

To summarise, the administration of CsA led to an increase of proliferation as expected and desired. But no changes at mRNA expression levels of NFATc target genes MYF5, p21, and UTRNA were detected which was unexpected (Chakkalakal et al., 2003). Hence, other signalling pathways are involved that participate in the induction of cell proliferation. Finally, deflazacort, which was suspected of being involved in the calcineurin/NFATc pathway, was tested using SkMCs.

4.4.4.3 Deflazacort

Up to now, only corticosteroids are routinely used with a slight beneficial outcome in DMD therapy (Biggar et al., 2006; Angelini, 2007). In mice, deflazacort was mentioned as positively influencing the calcineurin/NFATc pathway which led to an increased nuclear translocation of NFATc and UTRNA expression. This effect was diminished using CsA. (St-Pierre *et al.*, 2004) Several other mechanisms of action are also discussed including an anti-inflammatory effect. As modelled in the signalling network, it can be assumed that deflazacort decreases cell proliferation accompanied by increased expression of NFATc target genes MYF5, p21, and UTRNA.

Experiments using deflazacort were designed similarly to CsA (see section 4.4.4.2). Optimal concentrations of deflazacort resulting in an increase in proliferation and vitality are shown in table 40. It is noticeable that the optimal concentration of each patient and control individual varies in the molar range from µM to mM. A toxic concentration was not reached (>200mM).

table 40: Optimal concentrations of deflazacort after 48h treatment investigated in SkMCs using proliferation and vitality tests as well as expression studies.

	Cells	Deflazacort
Dys+	14/00	1mM
	18/01	10µM
Dys-	43/01	1mM
	77/02	1µM
	109/03	10µM
	145/03	1mM

The cell proliferation results of (Dys+) and (Dys-) SkMCs treated with deflazacort compared to DMSO-affected cells are shown in figure 27a. The data demonstrate an increase of cell proliferation for 3 of 4 DMD patient cells of 30-50% and for both controls of 21 to 61%. The (Dys-) SkMC 43/01 indicated a slight stimulation of proliferation. This positive impact on proliferation disappeared after another 24h incubation time as described for okadaic acid and CsA (sections 4.4.4.1 and 4.4.4.2).

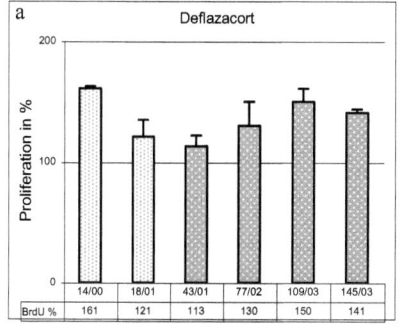 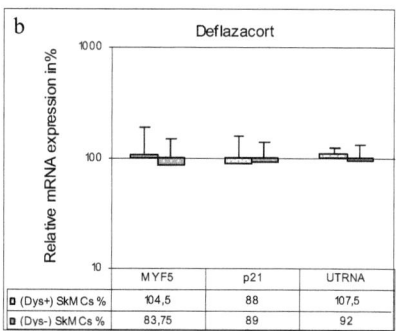

figure 27: (Dys+) and (Dys-) SkMCs treated with deflazacort for 48h.
a) The proliferation data were normalised to DMSO-affected SkMCs. The cells treated with the solvent DMSO are indicated by the 100% curve. b) The mRNA expression of MYF5, p21 and UTRNA was determined in myoblasts treated with the optimal concentration of deflazacort. The results were calculated relatively to DMSO-affected SkMCs (=100%).
(Dys+) SkMCs: 14/00; 18/01   (Dys-) SkMCs : 43/01; 77/02; 109/03; 145/03

The mRNA expression levels of NFATc target genes MYF5, p21, and UTRNA in myoblasts influenced by deflazacort are displayed in figure 27b. The SkMCs treated with DMSO are indicated by the 100% curve. In the calculation of the relative expression levels no differences were found using either DMSO-treated or untreated SkMCs for the comparison to deflazacort-treated SkMCs. For all 3 genes no alterations in the mRNA expression MYF5, p21, and UTRNA were determined.

In summary, the treatment of human SkMCs with deflazacort did not lead to an elevation of UTRNA expression as found in mice. An increase in proliferation was demonstrated for both (Dys+) and (Dys-) SkMCs, but was not a consequence of reduced p21 expression. The outcome is similar to okadaic acid and CsA, but it remains unclear which interference in NFATc signalling or other signalling pathways involved in the cell cycle are addressed by deflazacort.

A modification of NFATc target gene expression of MYF5, p21, and UTRNA could not be demonstrated for inhibition of calcineurin using okadaic acid and CsA. In this context, the consideration of deflazacort is restricted, since the mechanism of action of deflazacort is unknown. However, all the substances used exhibit a conspicuous tendency to trigger an elevation of cell proliferation when applied to human SkMCs as proposed by the Petri net model and the outcome of the Mauritius map analysis. So the question arose whether a combined administration of CsA and deflazacort would lead to an added increase in cell proliferation, and was investigated in subsequent experiments. For toxicity reasons okadaic acid was excluded.

RESULTS

4.4.4.4 Combination of cyclosporin A and deflazacort

Deflazacort and CsA are described as having a contrary impact on signal transduction pathways in skeletal muscle cells (St-Pierre *et al.*, 2004). However, a positive effect on DMD patients was noticed when a routine administration of deflazacort was combined with CsA (von Moers, 2007). Furthermore, a clinical study has been conducted using both CsA and prednisolone. Deflazacort is an oxazoline derivative of prednisolone with less severe side effects.

But, based on results from previous experiments, an increase in cell proliferation was expected at higher values than achieved with each substance accompanied by unchanged mRNA expression levels of the NFATc target genes MYF5, p21, and UTRNA.

The optimal concentrations of CsA and deflazacort were applied to SkMCs as a combination of both substances (table 41). An incubation time of 48h also turned out to be most sufficient as independently found for both compounds.

table 41: The concentrations of deflazacort and CsA used as combination in SkMCs. Proliferation and vitality tests as well as expression studies were performed after 48h treatment.

Cells		Deflazacort / CsA
Dys+	14/00	1mM / 5nM
	18/01	10µM / 10nM
Dys-	43/01	1mM / 10nM
	77/02	1µM / 50nM
	109/03	10µM / 1nM
	145/03	1mM / 50nM

The treatment of (Dys+) and (Dys-) SkMCs with both CsA and deflazacort resulted neither in the expected elevation nor in a reduction of cell proliferation (figure 28a). The enhancing impact as particularly found for CsA and deflazacort was diminished. A slight increase of cell proliferation was observed for 3 (Dys-) cell cultures and the (Dys+) SkMC 14/00, and was not a consequence of reduced p21 expression (figure 28b).

RESULTS

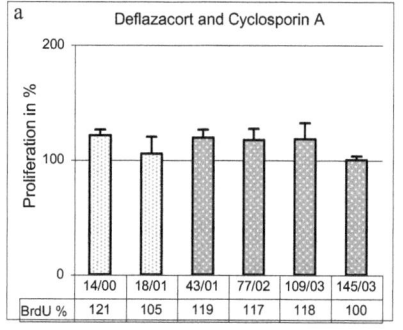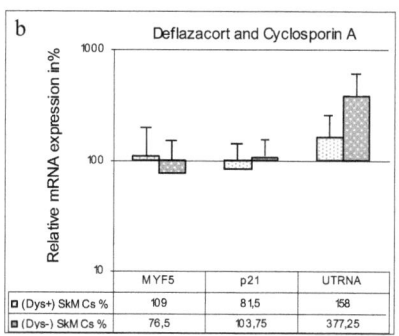

figure 28: (Dys+) and (Dys-) SkMCs treated with deflazacort and CsA for 48h.
a) The altered proliferation was normalised to DMSO-affected SkMCs. Cells applied to the solvent DMSO are indicated by the 100% curve. b) The mRNA expression of MYF5, p21 and UTRNA was determined in myoblasts treated with optimal concentrations of deflazacort and CsA. The results were calculated relatively to DMSO-affected SkMCs (=100%). Whereas the (Dys+) SkMCs were not altered by any of the drugs, the combination of both compounds shows a significant increase (P < 0.05) in UTRNA mRNA expression (3-7fold). The statistical analysis was performed using the students t-test.
(Dys+) SkMCs: 14/00; 18/01   (Dys-) SkMCs : 43/01; 77/02; 109/03; 145/03

The (Dys+) and (Dys-) SkMCs influenced by deflazacort and CsA were studied at mRNA level for NFATc target genes MYF5, p21, and UTRNA (figure 28b). The cells treated with DMSO are indicated by the 100% curve. The calculation of the relative mRNA expression of drug-treated SkMCs compared to DMSO-treated or untreated SkMCs produced equal results. Because of the restricted number of patients and controls, a relative difference of more than 200% and less than 50% was considered as an increase and decrease, respectively, in mRNA expression. For p21 and MYF5, the combination of both drugs exhibited no alteration in mRNA expression. In the (Dys+) SkMCs the mRNA expression of UTRNA remained unchanged (< 200%) as found for (Dys+) and (Dys-) myoblasts separately treated with the chemical substances. However, (Dys-) SkMCs treated with both deflazacort and CsA showed a significant increase ($P<0.05$) in UTRNA expression. An elevation of about 3- to 7fold was observed relative to DMSO-affected SkMCs. Since an increase at mRNA level only suggests an augmentation of the UTRNA protein expression, these findings were confirmed at protein level.

4.4.4.5 Expression analysis at protein level

The mRNA expression level does not necessarily correlate with the abundance at protein level. Regulation processes after transcription of the mRNA including splicing of the mRNA,

Results

transport into the cytosol, miRNA, half-life of the mature mRNA, and translation of the protein play a decisive role.

The mRNA expression of UTRNA was particularly increased in 2 DMD patients: 77/02 and 145/03. These SkMCs were chosen to verify the UTRN expression at protein level. To extract the proteins, the cells were grown in T25 flasks and treated for 72h. The optimal concentrations for each SkMCs cluture were used as determined before. The samples included in these experiments are specified in table 42.

table 42: The samples of all treated SkMCs and the respective drug concentration of 2 DMD patients (77/02 and 145/03) used for the detection of UTRN at protein level.

Treatment	77/02	145/03
Untreated	-	-
DMSO	7mM	7mM
Deflazacort	50nM	50nM
CsA	1µM	1mM
Deflazcort / CsA	50nM / 1µM	50nM / 1mM

The antibody against UTRN detects UTRNA and UTRNB. A specific antibody against UTRNA was not purchasable at the time experiments were performed. The western blot analyses were performed using 30µg and 40µg protein for 77/02 and 145/03, respectively. The house-keeping gene ß-actin was used to standardise the amount of protein used for each sample. The protein determination by the BCA method resulted in an average protein concentration of 1µg/µL. Although similar amounts of protein were used for each sample, the intensities of each band of ß-actin were different. The mRNA expression of UTRNA was increased for deflazacort in combination with CsA but not for the other samples. Thus, an elevated protein level was expected for deflazacort and CsA in comparison to DMSO-treated or untreated SkMCs. The results are representatively shown for DMD cell culture 77/02 in figure 29 A and B for UTRN and ß-actin, respectively.

RESULTS

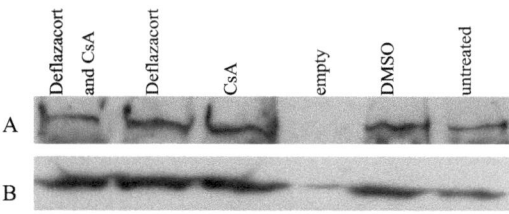

figure 29: The UTRN protein expression (A) was detected in treated and untreated (Dys-) SkMCs of the patient 77/02 by a monoclonal anti-UTRN antibody. As loading control, ß-actin (B) was used.

Since the intensities of the loading control vary between the samples, the differences in protein levels are not immediately determinable. Therefore a relative semi-quantitative determination of the UTRN protein expression of each sample was performed by comparing UTRN-specific signals with signals of the internal standard using the software BioDocAnalyzer 2.0 (Biometra). The results are shown in table 43.

table 43: Semiquantitative determination of UTRN protein levels in untreated and treated samples compared to DMSO-treated samples (100%). The house-keeping protein ß-actin was used as internal standard. The investigation was performed for the (Dys-) SkMCs 77/02 and 145/03.
* n.d. = not determined. Because of a gap in the protein band of the untreated cell sample 145/03, the band could not be analysed by the software.

Sample	77/02	145/03
untreated	101%	n.d.*
DMSO	100%	100%
CsA	102%	112%
Deflazacort	128%	229%
Deflazacort + CsA	167%	233%

The data show an increase in UTRN expression at mRNA as well as protein level. Interestingly, the treatment with deflazacort alone also resulted in elevated UTRN protein expression, which was not determined at mRNA level. For the validation, the mRNA expression data were verified using the same samples as employed for the protein analysis. Again increased mRNA UTRNA expression values were found for the samples treated with deflazacort and CsA but not with deflazacort or CsA alone. As found in previous experiments, mRNA levels of utrophin B are not detectable in human myoblasts, and are therefore not expected to impair the results.

5 Discussion

DMD is one of the most frequently inherited neuromuscular diseases in children. The disease is caused by mutation(s) in the *dystrophin* gene that result in a loss of the protein dystrophin. This loss is followed by the primary structural dysfunction and in addition it also influences several downstream processes. An efficient therapy is not available, but cell culture and animal model experiments as well as first clinical studies are in progress. These therapy strategies are based on:

a) external substitution of dystrophin by gene or myoblast transfer
b) reconstitution of patient's own dystrophin expression by skipping of mutated exons, trans-splicing, read-through of nonsense mutations
c) enhanced expression of dystrophin homologue protein UTRN by gene transfer or regulation of transcription factors
d) increase of patient's own muscle mass and/or muscle cell number by the use of gene technique such as transfection, RNAi or pharmaceutical substances
e) compensation of the dysfunction of processes downstream dystrophin.

(Bertoni, 2008; Scime and Rudnicki, 2008; Manzur *et al.*, 2008a)

The purpose of strategies (I) and (II) aim at substitution or production of dystrophin. Their effects are locally and temporally restricted and also involve the risk of provoking an immune response. In dystrophin-deficient skeletal muscle cells, a newly expressed dystrophin or its delivery system may be identified as neoantigen. An administration of immunosuppressant drugs is necessary.

The development of strategy alternatives that do not primarily concern dystrophin itself includes the attempt to compensate the loss of dystrophin [(III)-(V)]. In this context, the most interesting protein is the dystrophin homologue protein UTRN (III). Similarly to dystrophin, UTRN is able to link intra-cellular proteins of the DGC to the extracellular matrix but mainly at embryonic stages in contrast to dystrophin. Moreover, UTRN and dystrophin have different primary locations in muscle fibres. Whereas dystrophin is found throughout the length of muscle fibres, UTRN is primarily located at the postsynaptic membrane of the neuromuscular junction. The expression of the protein is up-regulated in DMD patients compared to normal individuals. In research studies an up-regulation of UTRN *in vivo* resulted in a rescue of the dystrophic pathology, but is not connected with muscle regeneration (Weir et al., 2004; Michel et al., 2007).

Regeneration of myoblasts (IV) was induced by several growth factors, e.g. myostatin inhibition (Bogdanovich et al., 2005; Nakatani et al., 2008), fibroblast growth factors (Neuhaus et al., 2003), the insulin-like growth factor (Schertzer et al., 2008), and calpain (Tidball and Spencer, 2000). First clinical approaches that also take into consideration these results in animal models or human cell culture include calpain inhibitors (Santhera Pharmaceuticals, Liestal, Switzerland) and myostatin repression (Wyeth-Ayerst, Madison, New Jersey, United States). Myostatin is a protein that is known to control skeletal muscle cell homeostasis. Human patients and animals suffering from a lack of myostatin show a remarkable muscle hypertrophy. The mdx mice treated with myostatin inhibitors show increased muscle mass and strength associated with an increase in fibre size (Manceau et al., 2008; Bradley et al., 2008). However, different opinions exist regarding muscle function and endurance. Whereas Bogdanovich et al. reported an improvement in muscle function and endurance (Bogdanovich et al., 2002), Qiao et al. discovered reduced endurance of treated mdx mice compared with untreated mice (Qiao et al., 2008). Results similar to Qiao et al. were previously found in cattle breeds that have double-muscle conditions caused by mutations in the myostatin gene. The cattle suffer from problems with fertility, calf viability, stress tolerance and endurance (Kambadur et al., 1997). A myostatin blockade does not produce an increase in myoblast proliferation and muscle fibre number. An augmentation of muscle mass by inhibition of myostatin therefore seems insufficient without an increase in skeletal muscle cell numbers.

Another strategy of DMD therapy is the enhancement of cell proliferation and differentiation including the reduction of fibrosis and necrosis of myoblasts [(IV)-(V)]. With regard of cell proliferation, the p21 expression was found to be significantly increased in skeletal muscle of DMD patients (Endesfelder et al., 2000). That provides a molecular explanation for reported considerable changes in proliferation and differentiation behaviour of (Dys-) skeletal muscle (Blau et al., 1983b). Application of ASO and siRNA in cell culture reduced p21 expression followed by an increase in cell proliferation (Endesfelder et al., 2003; Endesfelder et al., 2005). Interestingly, it was found that the p21 protein is consistently increased whenever the transcription factor cardiac ankyrin repeat protein (CARP) is up-regulated in mdx mice (Laure et al., 2009). A recent study reports on another protein, the fast myosin light chain 1 (MLCf1), that inhibits the transition from G1 to S phase of the cell cycle. The authors showed that the protein level of MLCf1 is increased in the diaphragm and gastrocnemimus and the cell proliferation is decreased. By contrast the extraocular muscle, which is less affected in DMD patients, had much lower level of MLCf1 and a higher proliferation rate (Zhang et al., 2008). That supports our strategy. But what still needs to be clarified using animal models and clinical studies is whether an increase in proliferation through a decrease of p21 expression has a positive impact on the function of skeletal muscle cells lacking dystrophin.

Regarding necrosis, the treatment with proteasome inhibitors resulted in a blockage of the proteasomal-dependent degradation pathway in mdx mice, and rescued the expression and membrane localisation of dystrophin associated proteins and led to a histological improvement of the dystrophic muscle (Spencer and Mellgren, 2002; Bonuccelli et al., 2007; Briguet et al., 2008). Muscle fibrosis was reduced by elevation of tissue inhibitor of matrix metalloproteinases expression at mRNA level (von Moers et al., 2005). A diminishment of muscle fibrosis and an increase in myoblast proliferation could ameliorate the dystrophic pathology of DMD patients in advanced cases.

The attempt to compensate the loss of dystrophin downstream dystrophin (V) relies on extensive expression studies on the one hand. These comparisons between DMD patients and normal controls are performed at RNA as well as protein level, and are partly based on chip technique (Bakay et al., 2002; Haslett et al., 2002; Haslett et al., 2003; Pescatori et al., 2007). For example, the cholesterol trafficking protein Nieman-Pick C1 is found to attenuate the dystrophic phenotype of the mdx mice and therefore suggested as potential therapeutic target for DMD (Steen et al., 2009). The expression of the delta-sarcoglycans at low level is proposed to attenuate the muscular dystrophy as studied in mdx mice (Li et al., 2009). On the other hand, it is also based on reports of DMD and BMD outlier patients which suggest that the lack of dystrophin might result in an altered activation of several signal transduction pathways in skeletal muscle (Tidball and Wehling-Henricks, 2004). In this context, it has been considered that dystrophin is part of a large complex of interacting proteins and that its absence is therefore associated with partly unknown pathophysiological downstream cascades.

Two siblings with an intra-familially different course of the disease were studied concerning differences at RNA level using a subtraction library that forms the basis of this work. A number of 5 genes was determined (CSNK1A1, RAP2B, DCTN3, CBFb, MYL2) which are increased and may trigger the mild phenotype of the younger DMD brother. All these studies supply information for several proteins of different signal transduction pathways. For most of them a link to dystrophin and the dystrophin downstream cascade is not self-evident.

DISCUSSION

The compilation of these processes and the simulation of different case scenarios *in silico* necessitate the application of a mathematical model. As a result of the lack of kinetic data the Petri net theory was chosen for the model development. This thesis firstly reports on a theoretical model connecting 2 main signal transduction pathways encompassing dystrophin. The network includes processes which are influenced by own experimental data, the loss of dystrophin downstream dystrophin and it considers parts of the cellular pathways which are important for the dystrophin signalling. However, it makes no claim to be complete. Analyses of this model resulted in experimental modulation of selected members of the network. The consequences of this modulation were studied at mRNA and protein level using human SkMCs in cell culture.

Using Petri net analysis techniques such as invariant and cluster analyses, the complex network was validated and decomposed in biologically meaningful sub-nets, which can be summarised in 2 main parts:

(1) dystrophin downstream part connecting dystrophin with p21 and UTRNA via phosphorylation of NFATc through CSNK1A1 and JNK1 and
(2) the RAP2B-calcineurin part leading to de-phosphorylation of NFATc.

Both parts have contrary influences on the transcription level of NFATc target genes.

5.1 Sub-net connecting dystrophin with p21 and UTRNA via phosphorylation of NFATc through CSNK1A1 and JNK1

5.1.1 Petri net analyses of the dystrophin downstream sub-net

This sub-net connects dystrophin with the DGC and the subsequent reaction cascade including the following main components: JNK1, c-Jun, CSNK1A1, and NFATc. It focuses on phosphorylation of NFATc through CSNK1A1. Although NFATc is the transcription factor of target genes, e.g. UTRNA, MYF5, and p21, its migration into the nucleus and transcriptional activity is disregarded since this sub-net only focuses on the inactivated phosphorylated form of NFATc. Therefore NFATc and its target genes are discussed in the de-phosphorylation sub-net in section 5.2.

DISCUSSION

Dystrophin initiates a complex downstream cascade, according to paper published by Oak et al. (Oak et al., 2003), but is extended after JNK1. The pathway leads to the phosphorylation of the transcription factors c-Jun and NFATc. Whereas c-Jun, a negative regulator of p21, is phosphorylated and activated by JNK1, the transcription factor NFATc can be phosphorylated and de-activated by JNK1 or CSNK1A1 and other kinases (Oak et al., 2003). CSNK1A1 is a member of the serine/threonine casein kinase I family (CKI), which contains another 6 isoforms encoded by distinct genes. The CK1 family phosphorylates numerous substrates, e.g. NFATc, p53, and myosin (Knippschild et al., 2005a). The NFATc target gene p21, the negative regulator of cell cycle signalling, implicates another pathway that includes the CDKs, the RB protein and its regulator E2F. This pathway is modelled in the signal transduction network to include proliferation deficits and the role of p21 in DMD patients. Considering this network and the subtraction library results, the following hypothesis was proposed:

In classic (Dys-) SkMCs, the phosphorylation of NFATc through JNK1 is decreased and associated with predominance of the de-phosphorylation pathway. This can consequently lead to an enhanced expression of NFATc target genes such as UTRNA, MYF5, and p21. It is supported by the DMD patient with the mild phenotype whose increased expression level of CSNK1A1 is proposed to compensate the decrease of JNK1.

The dystrophin downstream cascade and the resulting phosphorylation pathway were studied by invariant analyses and Mauritius maps. The latter indicate the impact, defined as a theoretical value, of a gene or protein on the network after its knockout. These investigations support the importance of CSNK1A1. A knockout of the de-activation of NFATc by CSNK1A1 has an impact of 45% on the network. Other knockout events concern dystrophin (37%) and the initiation of JNK1 expression (37%). Hence, the CSNK1A1 represents a most effective point of application to achieve a manipulation of NFATc activity and the expression of UTRNA, MYF5, and p21.

Another structural analysis, the invariant analysis, produced similar results. There is one exception that describes the t-invariant Inv55, which contains the cyclic pathway of phosphorylation (by CSNK1A1) and de-phosphorylation of NFATc, and combines both sub-nets. A knockout of CSNK1A1 would destroy this cyclic pathway, which represents a balanced ratio between phosphorylation and de-phosphorylation of NFATc. An increase in nuclear localisation of NFATc and the expression of the NFATc target genes are the consequence as hypothesised. The invariant and the Mauritius map results conform to expression data of CSNK1A1 of the subtraction library.

The refined invariant analysis through MCT-sets showed that the gene-regulation sub-nets modelled are in separate MCT-sets with regard to up-regulation or down-regulation of the gene. This is biologically meaningful and confirms the Petri net. The MCT-sets containing initiation of RB-protein-mediated cell cycle control are also separated concerning positive and negative cell cycle regulation and the different CDKs involved. This branching of the pathway led to a disconnected MCT-set as expected. Consequently, the Petri net and especially its Mauritius map analysis allow and, what is more, simplify the search for target proteins for the application of chemical substances as well as for candidate genes used in vector-DNA or siRNA experiments. Chemical compounds can influence pathways at protein level more easily than other techniques. However, the fact needs to be taken into consideration that chemical compounds may affect a wide range of target proteins rather than a specific protein. Vector-DNA and siRNA approaches are specific for one gene but do not directly have an effect on protein activities, which makes it reasonable to have a tool for a selection of those candidate genes which can be assumed to have the greatest impact on myoblasts of DMD patients.

5.1.2 RNA expression studies of selected components

The results of the subtraction library showed an increased mRNA expression level of CSNK1A1 in the DMD patient with the mild phenotype compared to that of the elder brother, a classic DMD patient. This indicated a potential role of CSNK1A1 for the mild DMD phenotype. This supposition was to be verified using 5 other classic DMD patients and 5 normal controls. Additionally further components of the pathway were determined.

Starting with the DGC, the phosphorylation of the transcription factors c-Jun and NFATc is a next major step. In (Dys-) patients, **c-Jun**, phosphorylated and activated by JNK1, is expected to be decreased compared to (Dys+) myoblasts. The mRNA expression level of c-Jun is not altered in all DMD patients. But its phosphorylation state is basically more pivotal for the sub-net than its intra-cellular abundance.

A similar situation is found for the kinase **JNK1**. Contrary studies are mentioned in the literature so that a definite assumption is difficult to formulate. Considering its function within the signal transduction pathway downstream dystrophin, decreased levels of JNK1 are expected in (Dys-) myoblasts that would predominate the activation pathway of NFATc and an increase of the expression of NFATc target genes UTRNA, MYF5, and p21. That is confirmed by the observation of a reduction of UTRN expression as soon as dystrophin expression appears in the adult skeletal muscle (Perkins and Davies, 2003; Miura and Jasmin, 2006). Nevertheless, one study reported the activation of JNK1 in association with elevated dystrophic pathology in mdx mice (Kolodziejczyk *et al.*, 2001). Contrary to this, skeletal muscle of physically exercised mdx mice showed aberrantly up-regulation of the phosphorylated form of JNK2, but not JNK1 (Nakamura *et al.*, 2005). Additionally, considering this sub-net an elevation of JNK1 may decrease expression of UTRNA, MYF5, and p21. The expression level of JNK1 was not altered in classic DMD patients what does not conform to expectation. A normal expression value was hypothesised for the mild DMD patient. Even so, the mRNA expression of JNK1 was significantly decreased in the mild phenotypic DMD patient ($P<0.05$) as expected for classic (Dys-) patients. However, it does not supply information about the activity of JNK1 within the patients, which is currently not determined in human DMD patients. Additionally, JNK1 is known to be involved in other signalling pathways such as the regeneration of neuronal processes, cell survival, and proliferation (Bode and Dong, 2007).

The same results are found for the target protein **NFATc**. But the transcription factor NFATc can be phosphorylated by several other kinases than JNK1, such as the kinase CSNK1A1, which was up-regulated in the mild DMD patient compared to the classic DMD brother. Similar to c-Jun and JNK1, the phosphorylation state is likely to be more important than the expression level.

A completely different situation is shown for **CSNK1A1**. An increase in CSNK1A1 expression was previously found in the mild DMD patient compared to the elder brother, and it was assumed that CSNK1A1 may be responsible for the mild phenotype. For that reason, an increased expression was also expected for the mild phenotype compared to the other 5 classic DMD patients. But the expression of CSNK1A1 in classic DMD patients are significantly increased compared to controls ($P<0.01$). By contrast in the mild DMD patient the value is significantly decreased ($P<0.05$) in comparison to classic DMD patients. The expression results of the DMD patient with the mild clinical course are similar to the normal (Dys+) situation and are consistent with the hypothesis. However, the mRNA expression level does not necessarily correlate with the protein level or its phosphorylation status, which are not determined in human myoblasts.

To summarise, the kinase CSNK1A1 is a very interesting candidate for further studies because of its association with DMD pathology. However, what needs to be examined is whether an increase in CSNK1A1 expression exacerbates DMD pathology and its decrease induces a mild phenotype. With reference to the data of the intra-familial comparison, it can be assumed that this is not the case. But this relation between the 2 siblings is in contrast to the comparison with 5 DMD patients. Their CSNK1A1 expression level is significantly increased contrary to the mild DMD patient. This discrepancy is to be examined and forms the basis for further investigations.

5.1.3 Experimental modulation of CSNK1A1 expression

For all applications, an increase of proliferation should ideally be accompanied by with an unchanged or elevated UTRN expression. In this context, the deactivator CSNK1A1 of the transcription factor NFATc, is studied and its impact on proliferation and transcription of NFATc target genes, e.g. UTRNA, MYF5, and p21, using SkMCs in cell culture.

The siRNA technique allows a specific interference in the CSNK1A1 expression in cell culture, but not in its activity. The mRNA expression level of CSNK1A1 persisted at normal level in (Dys+) SkMCs. It seems reasonable to assume that (Dys+) myoblasts are able to diminish siRNA consequences efficiently. This supposition is supported by the fact that the CSNK1A1 expression was reduced in (Dys-) SkMCs (from 100% down to 8-45%). An efficient counter-regulation might be absent in (Dys-) myoblasts. The expression levels of UTRNA, MYF5, and p21 remained unchanged. The expression level of NFATc was not determined since siRNA against CSNK1A1 may change the phosphorylation state of NFATc but not its expression level. In summary, down-regulation or inhibition of CSNK1A1 did not show the expected effects.

Up-regulation experiments using cDNA plasmid vectors were also performed since the DMD patient with the mild phenotype showed enhanced mRNA expression of CSNK1A1 compared to the brother with a typical clinical course of DMD. Considering the signal transduction pathway, an up-regulation of CSNK1A1 leads to an enhanced accumulation of cytosolic NFATc which consequently reduces the expression of NFATc target genes and increases proliferation. The effect of CSNK1A1 up-regulation (1,5-82fold) was dependent on the DMD patient so that the results did not produce a consistent pattern. The proliferation of SkMCs was reduced in all DMD patients, which is inconsistent with the expectation. Except for

DISCUSSION

2 DMD patients, the expression levels of UTRNA, MYF5, and p21 were enhanced or remained unchanged.

Taken together, up- and down-regulation of CSNK1A1 in (Dys-) SkMCs succeeded. The results show that neither down- nor up-regulation of CSNK1A1 could influence the NFATc target genes as expected and modelled in the signal transduction pathway. The inconsistency in the results may be due to the following factors. It is possible that the skeletal myoblasts in culture react differently to treatments compared to cells in their original tissue. The muscle tissue of DMD patients and controls were taken at different ages and consequently at different stages of the disease and muscle development. All the factors mentioned can have an important impact on the effect of any treatment and can lead to individual variations.

Furthermore CSNK1A1 is primarily involved in other signalling pathways as described above, which do not affect the parameters determined. In the literature, JNK1 and glycogen synthase kinase-3 (GSK3) are also mentioned as kinases for NFATc (Beals *et al.*, 1997; Dong *et al.*, 1998), but none of these is stated as the primary kinase of NFATc. Otherwise, NFATc might not be the most important transcription factor for UTRNA, MYF5, and p21. For example, the transcription factor NFATc is particularly known to regulate the production of interleukins in T-cells, eg. IL-4, which activates and increases the immune response of T-lymphocytes (Yoshida *et al.*, 1998). However, in mdx mice it was shown that the expression of UTRNA is regulable via the NFATc pathway (St-Pierre *et al.*, 2004), but no experiments concerning the phosphorylation by CSNK1A1 have so far been performed. The antagonist of NFATc phosphorylation is its de-phosphorylation by calcineurin, which is stated as the primary phosphatase of NFATc in the literature (Crabtree and Olson, 2002), and is part of the second sub-net of the signal transduction pathway. With regard to the hypothesis and the results obtained a clear conclusion cannot be drawn.

5.2 Subnet connecting de-phosphorylation of NFATc via RAP2B-triggered calcineurin activation

5.2.1 Petri net analyses of the RAP2B-NFATc sub-net

This sub-net connects the RAP2B-mediated calcium release that activates calcineurin and contains the following main components: RAP2B, calcineurin, NFATc, UTRNA, MYF5, and p21.

Among other proteins, RAP2B mediates calcium release from the ER via PLCe and IP3 pathway. A significant role of IP3Rs in spontaneous calcium release was observed in dystrophin-deficient cells (Balghi *et al.*, 2006). Abnormal calcium homeostasis is observed in DMD muscle (Constantin *et al.*, 2006). Transfection of BMD mini-dystrophin in a dystrophin-deficient mouse cell-line reduced IP3R-mediated calcium release activity, and provides evidence for regulation of an IP3 pathway by mini-dystrophin (Balghi *et al.*, 2006). Via IP3R, calcium is released into the cytosol. The inflowing calcium is primarily captured by either troponin C or calmodulin which activates both proteins.

Important intra-cellular functions of calcium include muscle contraction and motility. In 1996, Anderson et al. studied binding of active calmodulin to the carboxy-terminal domain of dystrophin (Anderson *et al.*, 1996). The presence of calmodulin increases, via further enzymes such as the protein kinase II, the rate of phosphorylation of dystrophin that was shown for several isoforms of dystrophin (Madhavan and Jarrett, 1994; Calderilla-Barbosa *et al.*, 2006). However, a negative feedback to the calmodulin pathway through the lack of dystrophin is not to be expected. It is known that calmodulin activates kinases which can phosphorylate dystrophin, e.g. protein kinase II (Madhavan and Jarrett, 1999). This process is upstream dystrophin. This calcium binding protein has several interaction partners, e.g. kinases and phosphatases, which in turn interact with various target proteins (Ishida and Vogel, 2006). For this study it was proposed that the loss of dystrophin should not have an effect on calmodulin or its activity in that way and consequently on the processes considered in this network. Due to clarity, modelling of this calmodulin-dystrophin interaction has been omitted. However, the inhibition of active calmodulin signalling impairs the dystrophic pathology as demonstrated in mdx mice (Chakkalakal *et al.*, 2006).

Calmodulin combined with calcium activates the phosphatase calcineurin, formerly known as serine/threonine protein phosphatase 2B. Calcineurin activation is in particular required in slow muscle fibres. The activation of calcineurin can also be triggered through the proteolytic cleavage of calcineurin by calpain, which is also regulated by calcium fluxes (Crabtree and Olson, 2002; Wu *et al.*, 2007). Several chemical substances, e.g. CsA, are known to intervene in the activation processes of calcineurin. Calcineurin-mediated de-phosphorylation of NFATc unmasks a nuclear localisation sequence that allows the migration of NFATc into the nucleus and coupling to various other transcription factors, e.g. MEF2, AP-1 (Fos/Jun), GATA, Sp1, SP3, and HDAC (Im and Rao, 2004).

Target genes vary depending on cell type. In skeletal muscle, NFATc aims at several genes that especially encode slower, more oxidative muscle proteins and other proteins involved in muscle regeneration and differentiation, e.g. MYF5, p21, UTRNA (Michel *et al.*, 2007). In turn, inhibition of calcineurin/NFATc signalling is shown to block muscle fibre hypertrophy (Dunn *et al.*, 2002), induces slow-to-fast phenotype transformations (Serrano *et al.*, 2001), and suppresses UTRNA expression (Chakkalakal *et al.*, 2003). This is consistent with a study of Chakkalakal et al., which determined increased UTRNA expression in slower muscle fibres compared to fast muscle fibres (Chakkalakal *et al.*, 2003).

Taking into account this sub-net and the subtraction library results, the following hypothesis was established, and describes the opposed signalling pathway of NFATc activity as drawn for dystrophin downstream sub-net:

The level of cytosolic calcium is increased in DMD patients and this leads to an elevated activity of the phosphatase calcineurin. The augmented de-phosphorylation of NFATc by calcineurin produces an enhanced expression of NFATc target genes, e.g. UTRNA, MYF5, and p21. The elevation of MYF5 and p21 expression negatively influences the DMD disease through decrease of myoblast proliferation. The increase in UTRNA expression is suggested to ameliorate the dystrophic pathology, but is inefficient in the absence of proliferation and the lack of myoblasts. Consequently, a therapy strategy should focus on the elevation of myoblast proliferation. The expression of UTRNA ought to be at least at unaltered levels. Considering the signal transduction pathway modelled this postulation should not be achievable.

Using Mauritius map and invariant analyses, the RAP2B-calcineurin pathway turned out to be the part most relevant for the net behaviour. A knockout in that sub-net would lead to a malfunction of 78% of the whole Petri net model. In the literature, this hypothetical effect is demonstrated by several knockout animal models. For example, the homozygous calreticulin knockout mouse is lethal at the embryonic stage because of failing in heart development

(Mesaeli et al., 1999). A mutation in PLCe leading to malfunction of the protein is implicated with the nephrotic syndrome in humans (Hinkes et al., 2006). The inhibition of calcineurin by CsA and FK506 causes profound bone loss in animal models suggesting a role of calcineurin in skeletal remodelling (Sun et al., 2005). Therefore, calcineurin, the phosphatase of the key component NFATc, is assumed to have an essential impact on NFATc-mediated transcription, as for example on the increase of p21 transcription as one of its favourite target genes. The latter can enhance the decrease of myoblast proliferation.

5.2.2 RNA expression studies of selected components of the RAP2B downstream cascade

Using the subtraction library, the mRNA expression level of RAP2B was increased in the mild DMD patient compared to the classic DMD brother. A comparison of the mild DMD patient with 5 other classic DMD patients and 5 normal controls is used to verify this result and again further components of the pathway were analysed.

In (Dys-) patients, **RAP2B** expression is expected to be higher than in (Dys+) controls to regulate calcium fluxes. An increase in RAP2B expression is proposed to cause high cytosolic calcium concentrations. In DMD patients the calcium balance between the cytosol and ER has shifted to the cytosol. This is consistent with increased expression levels of RAP2B in classic DMD patients. By contrast, the mild phenotypic DMD patient shows normal RAP2B expression data. Therefore, it can be concluded that the elder brother patient may not be representative of classic DMD patients, as confirmed by the results of CSNK1A1 mRNA analysis (see section 5.1.2). A high cytosolic calcium concentration as found in the classic DMD patients may lead to an elevated requirement of **calcineurin**. However, the expression levels of all DMD patients are comparable to normal controls. In addition, the increased calcium concentration produces calcineurin activity and does not necessarily induce calcineurin expression.

The same is also applicable to **NFATc**. Elevated calcineurin activity is proposed to cause an enhanced expression of NFATc. In comparison to normal controls, the mRNA values of NFATc were unchanged in classic DMD patients, but the mild DMD patient shows significantly decreased values ($P<0.05$). However, the phosphorylation state of NFATc is decisive. Calcineurin-mediated de-phosphorylation of cytosolic NFATc leads to increased nuclear translocation of NFATc and augmented transcription of NFATc target genes UTRNA,

MYF5, and p21. On the basis of the known calcium homeostasis in DMD patients, elevated NFATc activity is to be expected even though the mRNA values of NFATc remain unchanged. Consequently, the expression level of UTRNA, MYF5, and p21 are expected to be increased in DMD patients. Interestingly, the lowest levels of the **p21** and **MYF5** expression are determined in the patient with the mild phenotype.

Furthermore, the DMD patient with the mild phenotype also shows the lowest **UTRNA** data, which indicates a reduced significance of UTRNA in muscle regeneration (Weir *et al.*, 2004). Nevertheless, a high expression level of UTRNA is generally assumed in the literature. From that point of view it seems useful to modify NFATc activity as suggested by Chakkalakal et al. (Chakkalakal *et al.*, 2003) as a therapy strategy for up-regulation of UTRNA, the homologous protein of dystrophin. But 3 aspects have additionally to be taken into account for such a strategy. The UTRNA mRNA level is already increased in all DMD patients. In these patients no positive correlation exists between the phenotype and the UTRNA value that supports the results of this study. Data from mdx mice suggest an approximately 4fold rise in UTRNA protein levels (Weir *et al.*, 2004). For mdx/CnA* (mdx mice expressing enhanced muscle calcineurin activity) a 2fold increase is described, which already attenuates dystrophic pathology (Chakkalakal *et al.*, 2004). This phenomenon in mice is obviously different from humans.

In addition, not only UTRNA but also p21 and MYF5 have NFATc as a transcription factor. Almost the same pattern is found for Sp1, which was discussed as a potential transcription factor to increase UTRNA expression via okadaic acid (Rodova *et al.*, 2004), but neglects 6 Sp1 sites in the p21 gene (Lu *et al.*, 2000). Consequently, an increase in NFATc expression or activity could imply an up-regulation of p21 and MYF5. Both are already elevated in the majority of DMD patients and reduce proliferation and muscle regeneration. This effect is not conducive to delaying the progression of muscular dystrophy. Concerning a strategy downstream of dystrophin to slow down the muscle dystrophy process, a regulated reduction of p21 (Endesfelder *et al.*, 2003; Endesfelder *et al.*, 2005) and MYF5 is beneficial in combination with unaltered or elevated UTRNA expression levels. As a result, both myoblast proliferation and the dystrophin homologous protein are enhanced at the same time. In consideration of the Petri net this strategy does not seem feasible. But, contrary experimental results in DMD patients, e.g. the failed modulation of NFATc target genes by artificially modified CSNK1A1 expression, indicate a more complex signal transduction pathway than drawn in the Petri net.

Consequently, it is not impossible that the favoured reduction of p21 and MYF5 can be combined with an elevation or unchange of UTRNA expression through regulation of NFATc. To answer this question, attention should be given to chemical substances already used in medical applications which influence the calcineurin/NFATc signalling pathway.

5.2.3 Modulation of calcineurin activity using pharmacological interventions

Chemical calcineurin inhibitors include CsA, FK506 (tacrolimus), and okadaic acid. Okadaic acid inhibits calcineurin to a lesser extent than the protein phosphatases 1 and 2A. However, this toxin is proposed to induce UTRN expression (Rodova et al., 2004), and it is that it influences the calcineurin/NFAT signalling. The precise adjustment of the okadaic acid concentration for each DMD patient and control SkMCs resulted in an increase of cell proliferation in some SkMCs cultures. However, this concentration is very close to its high toxic concentration. Okadaic acid is non-applicable in medical treatment. CsA and FK506 have similar modes of action. But, FK506 has more severe side effects, and is suspected of causing various types of cancer. Consequently, FK506 and okadaic acid were not considered in the study.

The calcineurin inhibitor CsA is controversially reported in studies using mdx mice and in human medical applications. Stupka et al. suggest that the active calcium dependent calcineurin pathway is responsible for the benign phenotype of mdx mice compared to DMD patients, and definitely negates a positive effect of CsA in mdx mice (Stupka et al., 2004). By contrast, another study in mdx mice demonstrated improved muscle function after administration of CsA for 4-8 weeks and exercise on treadmill without changes in the calcium homeostasis (De et al., 2005). Even so, the mode of action of CsA and calcineurin are not yet fully understood and may be different in the human dystrophic pathology.

To our knowledge, this study is the first to conduct a treatment of CsA in SkMCs of DMD patients and controls. As modelled in the signal transduction pathway, the inhibition of calcineurin leads to a reduced expression of NFATc target genes, UTRNA, MYF5, and p21. That is expected to improve cell proliferation. In this study the proliferation rate is elevated in SkMCs (>10-47%), and tends to be more increased in (Dys-) SkMCs.

Improved cell proliferation can be initiated in (Dys-) SkMCs by down-regulation of the expression level of p21 as demonstrated in this group (Endesfelder et al., 2005). The expression levels of UTRNA, MYF5, and p21 showed neither an up-regulation nor a down-regulation compared to DMSO-treated SkMCs. The expression data of UTRNA, which remained unchanged, are in line with earlier observations (De *et al.*, 2005). Consequently, CsA-governed proliferation uses a different signalling pathway that may not involve a reduction of p21 expression. Therefore various proteins/pathways can be considered, e.g. the cyclin-dependent kinases, the regulation of synthesis of S-phase genes including RB protein and E2F, and the transcription factor p53 that regulates proteins of the cell cycle and apoptosis.

The glucocorticoid deflazacort is thought to activate calcineurin/NFAT signalling, and, among other pharmacological approaches, can slow down the progression of DMD (Biggar *et al.*, 2006; Angelini, 2007). Another glucocorticoid used in DMD therapy is prednison, but deflazacort produces fewer side effects than prednison, and was therefore preferred in this work. Although a therapeutic effect is visible, the molecular basis of the pharmacological effect is not well understood. Investigations in mdx mice indicated an activation of the calcineurin phosphatase associated with an up-regulation of the NFATc target gene UTRN using deflazacort (St-Pierre *et al.*, 2004). A very recent study reports an ameliorated preservation of the spin-spin relaxation time in some muscle types in DMD patients treated with deflazacort (Mavrogeni *et al.*, 2009).

In this study for the glucocorticoid deflazacort, an increase in cell proliferation was investigated for (Dys+) SkMCs and (Dys-) SkMCs (>13-50%). But, no alteration in mRNA expression of UTRNA, MYF5, and p21 was determined which in this case excludes p21 as a potential mediator of the raised cell proliferation. Since this study shows improved cell proliferation for deflazacort and CsA, the question arises whether the positive impacts of both pharmacological applications can be added up by using a combination of both substances. But, as described in the literature, the modes of action of both are in fact proposed to function in the opposite direction. However, not only deflazacort but also CsA were already suggested and used as drugs in DMD therapy, even though with different results (Sharma *et al.*, 1993; St-Pierre *et al.*, 2004; Stupka *et al.*, 2004). Additionally, the German Society for Muscle Diseased Persons presently performs a clinical trial using a combination of prednison and CsA (Korinthenberg, 2008).

Consequently, a combination of deflazacort and CsA was tested. In this study, the combination of CsA and deflazacort reduces or impairs the elevated proliferation of the SkMCs as separately shown for CsA or deflazacort. But the combination of deflazacort and CsA produces a significant elevation of UTRNA mRNA levels ($P<0.05$) and increased protein expression compared to either drug alone. Because of the unchanged mRNA expression level of UTRNA after treatment with deflazacort or CsA, it can be assumed that NFATc might not be the activator of the UTRNA expression. Several other transcription factors, e.g. Sp1, GABP alpha/beta, and the artificial 4-zinc-finger protein, Bagly, as well as the myogenic Akt signalling are also feasible (Gyrd-Hansen *et al.*, 2002; Onori *et al.*, 2007; Peter *et al.*, 2009). The 3- to 7fold increase in UTRNA mRNA expression is followed by an elevated protein level (1.6-2.3fold) in human SkMCs without decreasing cell proliferation. Such an increase in UTRNA expression is consistent with the results recently reported using an artificial transcription factor called Vp16-Jazz (Mattei *et al.*, 2007; Desantis *et al.*, 2009).

Taken together neither de-phosphorylation nor phosphorylation of NFATc have an influence on the expression level of the target genes UTRNA, MYF5, and p21. Therefore, NFATc may not be crucial as a transcription factor with an important impact on the expression levels of these genes and on p21 regulated proliferation (St-Pierre *et al.*, 2004). This part of the hypothesis was not supported by the results. As seen in previous experiments, the expression of NFATc also remained unchanged. Nevertheless, an increase in proliferation is determined for CsA and deflazacort for the majority of the SkMCs tested. Other modes of action which may lead to enhanced cell proliferation should be discussed for deflazacort and CsA as well as other NFATc target genes.

(1) Some chemical substances, such as glucocorticoids, are assumed to be able to integrate into the DNA (von Knebel *et al.*, 1991), either to influence the histone code or to simulate or initiate transcription factor functions. That could be a potential explanation for deflazacort but seems to be unlikely for CsA. (2) The inhibition of calcineurin and the subsequent cytosolic localisation of NFATc may involve more complex biochemical consequences than those modelled in the Petri net and stated in the literature. (3) The age and/or stage of a DMD patient might also have an influence on the effect of pharmacological interventions.

But, contrary to expectations as proposed in the hypothesis, it is possible to influence target genes of NFATc at different levels. Increased UTRNA expression values are found using a combined application of CsA and deflazacort and this does not lead to an elevation of p21 expression followed by a diminished proliferation.

The question arises whether this increase in UTRNA expression in cell culture suffices to improve dystrophic pathology or to slow down the progression of DMD patients. It therefore remains to be seen whether the data of the clinical trial of the drug combination produce results as promising as the one explored in this study.

5.3 Future directions

Future directions include investigations in systems biology and molecular biology with the two fields collaborating closely.

In systems biology, it is intended to refine the Petri net model, but also to extend the modelled pathway which particularly includes transcription factors, further NFATc target genes and cell cycle proteins. This extension will lead to a very complex model that is difficult to overview if modelled at one hierarchical level. Therefore, a hierarchical Petri net with 2 or more levels must be considered in which, for example, the transcriptional level can be modelled into one hierarchical level. The discovery of other transcription factors and connecting pathways will also make it possible for analyses, such as Mauritius maps, to identify other points of application for chemical substances and modification of gene expression and thereby therapy strategies downstream dystrophin.

Consequently, in molecular biology, it is necessary to discover specific transcription factors or artificial molecules. Both make it easier to enhance or to silence expression of each of the relevant genes separately. In the context of detected molecules, which specifically interact with DNA binding domains, the search for NFATc inhibitors using *in vitro* selection technology (SELEX) on combinatorial RNA libraries with random nucleotide sequences should be mentioned (Bae *et al.*, 2002; Cho *et al.*, 2004). Alternatively, attempts to increase selectivity of gene expression by transcription factors, gene substitution and silencing for example via RNA-interference could be performed in parallel. Several factors come into consideration. Dogra et al. report increased activation of a number of about 30 nuclear transcription factors in mdx mice that are involved in numerous cell signalling pathways (Dogra *et al.*, 2008). The CARP protein is one transcription factor that is increased in the mdx model (Laure *et al.*, 2009). That can be applied for modelling and animation of a hierarchical Petri net representing the DMD pathology and the normal situation.

A very important further development would be an increase in the number of DMD patients at different ages of biopsy and control SkMCs to corroborate the statistical significance of the data. Using human primary myoblasts, an increase as well as a knockdown of further several proteins of the net is intended to validate the network itself and its analyses results. The subsequent examinations at mRNA and protein level complement the modelling of expression regulation. However, the analyses used are very specific to one gene or protein. Modulation of proteins or gene expression has rather a complex impact on several proteins and subsequent pathways. For this application antibody microarray analysis may be used in future experiments, in particular to compare treated and untreated SkMCs. This is useful for studying the role of pharmacological interventions in (Dys+) SkMCs compared with (Dys-) SkMCs. Clontech offers a comprehensive antibody microarray which can detect approximately 500 proteins covering protein kinases, cell cycle, neurobiology, cancer, and apoptosis. Many proteins of the Petri net model are included in this array (Andersson *et al.*, 2005; Chaga, 2008). Furthermore, not only the protein abundance but also activity tests of proteins, in particular concerning de- and phosphorylation, are of special interest since many protein interactions of the Petri net model are based on these protein modifications. For this application, conventional ELISA tests and Western blots can be considered as well as protein interaction assays using chip technology. Wide scale protein expression and activity analyses mediate the understanding of the molecular basis of DMD and its effective pharmacological and gene modifying interventions.

The iterative process of theoretical modelling and molecular biological experiments will give more insight into the biological pathways and consequently can advances the understanding of the pathogenesis of various diseases.

6 Summary

Duchenne muscular dystrophy (DMD) is one of the most frequently inherited neuromuscular diseases in children. It is caused by the absence of dystrophin (Dys-), what influences several downstream signal transduction pathways. To date no successful therapy is available. Currently, strategies are being considered which involve dystrophin downstream processes to make up for the lack of dystrophin. Analysing 2 siblings with an intra-familially different course of DMD using a cDNA subtraction library, 5 genes were determined to be up-regulated in the brother with the mild phenotype that might be responsible for a compensatory effect (Sifringer et al., 2004).

In this thesis, extensive data base searches associated 2 of these genes, ras related protein 2B (RAP2B) and casein kinase 1 A1 (CSNK1A1), to signal transduction pathways downstream dystrophin with de-/phosphorylation of the transcription factor nuclear factor of activated T-cells (NFATc) as intersection point. The established network can be sub-divided into to sub-nets. The first sub-net, the RAP2B downstream pathway, activates NFATc by its de-phosphorylation via the phosphatase calcineurin. The second sub-net, the dystrophin downstream cascade, represents the de-activation pathway of NFATc by its phosphorylation through CSNK1A1 and the c-Jun N-terminal kinase 1 (JNK1). It first represents the complex processes of DMD pathology in a Petri network which is a tool of systems biology for modelling of discrete distributed systems at different abstraction levels that can be applied to gene regulatory processes. Theoretical knockout analyses of the Petri net model using Mauritius maps and invariants identified de-/activation of NFATc as the most relevant part of the network.

For further investigations, in addition to RAP2B and CSNK1A1 7 candidate genes were selected from the de- and the phosphorylation part of NFATc: calcineurin, JNK1, jun oncogene (c-Jun), myogenic factor 5 (MYF5), the cyclin-dependent kinase inhibitor 1A (p21), NFATc, and utrophin A (UTRNA). UTRNA is homologue to dystrophin. MYF5 and p21 regulate differentiation and proliferation. All 3 are positively regulated target genes of the transcriptions factor NFATc. The transcription factor c-Jun negatively regulates p21 expression. The mRNA expression levels of these genes were studied in 5 classic DMD patients, the mild DMD patient, and 5 normal controls using Real-Time PCR. In most cases no significant differences were detectable. But contrary to the results of the subtraction library, which are restricted to 2 patients, the mild DMD patient represents a significantly decreased CSNK1A1 mRNA level in comparison with the classic DMD patients ($P<0.05$). The same results were found for NFATc and JNK1 ($P<0.05$) in comparison to (Dys+) individuals. Moreover, the lowest levels of the p21 and MYF5 expression were also determined in the phenotypically mild patient. The UTRNA gene is expressed at nearly normal level compared to classic DMD patients with increased values of up to 14fold. These results give a preference to an increase of proliferation to slow down the muscle dystrophy process that should ideally coincide with an unchanged or elevated UTRNA expression for appropriate therapeutic concepts, which currently concentrate mainly on UTRNA.

It was to be tested whether the p21 regulated myoblast proliferation and the increase in UTRNA protein level can be enhanced at the same time even if both proteins have NFATc as transcription factor. The contradictory effect of CSNK1A1 in DMD patients and the most interesting position of calcineurin within the network as antagonist of CSNK1A1 led to further investigations concerning the modulation of both proteins and their effects on proliferation and gene expression of NFATc target genes in human skeletal muscle cells (SkMCs) in culture. Considering the network model, an increase of CSNK1A1 expression is proposed to reduce the expression of UTRNA, MYF5, and p21 followed by elevated SkMCs proliferation. The gene expression of CSNK1A1 was successfully altered by plasmid cDNA vector and siRNA transfection. But no clear effect was found on proliferation and gene expression level of UTRNA, MYF5, and p21. Calcineurin was modulated by deflazacort and cyclosporin A (CsA). The separate use of deflazacort and CsA led to an increased proliferation of SkMCs without affecting the expression of MYF5, and p21 at mRNA level. The combined application as suggested in a clinical study showed slightly elevated proliferation results. The UTRNA mRNA and protein expression was not changed by CsA. But enhanced values were observed when using deflazacort. This effect was increased by the concurrent application of both substances. At the mRNA level an increase of about 3- to 7fold was observed and of about 2fold at protein level. The effect on UTRNA expression seems to be restricted to SkMCs of DMD patients.

To summarise, neither de-phosphorylation nor phosphorylation of NFATc has a clear influence on the expression level of the target genes UTRNA, MYF5, and p21. This suggests that NFATc may not be the important transcription factor for the target genes and consequently on p21 regulated cell proliferation. But an elevation of proliferation was observed for (Dys-) SkMCs by application of deflazacort and/or CsA which indicates other target points than NFATc. In contrast to the expectations from the Petri net the proliferation was elevated without a concurrent decrease of UTRNA expression, which gives positive impetus for appropriate therapy strategies but shows that the Petri net will have to be extended.

7 Zusammenfassung

Die Muskeldystrophie Duchenne ist eine der am häufigsten vererbbaren neuromuskulären Erkrankungen im Kindesalter. Es wird ausgelöst durch das Fehlen des Proteins Dystrophin (Dys-), welches in eine Vielzahl von Signaltransduktionswegen involviert ist. Derzeit steht keine erfolgreiche Therapie zur Verfügung. Gegenwärtig werden Strategien in Betracht gezogen, welche Dystrophin nachgeschaltete Prozesse einbeziehen, um den Verlust von Dystrophin zu kompensieren. Durch die Transcriptomanalyse eines Geschwisterpaares mit einem intra-familiär unterschiedlichen Verlauf der DMD wurden fünf Gene mittels einer cDNA Subtraktionsbank identifiziert, die in dem Bruder mit dem milden DMD-Phänotyp hochreguliert sind und damit für einen kompensatorischen Effekt verantwortlich sein könnten (Sifringer et al., 2004).

Mit Hilfe von umfangreichen Datenbankrecherchen konnten zwei dieser Gene, das Ras Related Protein 2B (RAP2B) und die Casein Kinase 1 AI (CSNK1A1), in Zusammenhang zu Dystrophin nachgeschalteten Signaltransduktionswegen gestellt werden, deren Knotenpunkt die De-/Phosphorylierung des Transkriptionsfaktors Nuclear Factor of Activated T-cells (NFATc) darstellt. Das dabei etablierte Netzwerk kann in zwei Teilnetze unterteilt werden. Das erste Teilnetz, der RAP2B nachgeschaltete Signalweg, stellt die Aktivierung von NFATc durch dessen De-Phosphorylierung über die Phosphatase Calcineurin dar. Das zweite Teilnetz, die Dystrophin nachgeschaltete Reaktionskaskade, beinhaltet den De-Aktivierungsweg von NFATc durch Phosphorylierung mittels CSNK1A1 und der c-Jun terminalen Kinase 1 (JNK1). Beide veranschaulichen erstmalig die komplexen Prozesse der DMD-Pathogenese in einem Petri-Netz. Ein Petri-Netz ist ein mathematisches Modell der Systembiologie zur Modellierung von diskreten Systemen auf unterschiedlichen Abstraktionsebenen und findet bei genregulatorischen Prozessen Anwendung. Theoretische Knockout-Analysen des Petri-Netzes mit Hilfen von Mauritius-maps und Invariant-Untersuchungen ergaben, dass die De-/Aktivierung von NFATc das bedeutendste Element des Netzwerkes ist.

Für weitere Studien wurden, zusätzlich zu RAP2B und CSNK1A1, sieben weitere Kandidatengene aus dem De- und dem Phosphorylierungs-Netz von NFATc ausgewählt: Calcineurin, JNK1, Jun Oncogene (c-Jun), Myogenic Factor 5 (MYF5), Cyclin-Dependent Kinase Inhibitor (p21), NFATc und Utrophin A (UTRNA). Utrophin ist homolog zu Dystrophin. MYF5 und p21 regulieren die Differenzierung bzw. Proliferation von Muskelzellen. Alle drei sind positiv regulierte Zielgene von NFATc. Der Transkriptionsfaktor c-Jun hingegen beeinflusst die p21-Expression negativ. Das mRNA-Expression-Niveau von diesen Genen wurde in fünf klassischen DMD Patienten, dem milden DMD Patienten und fünf Normalkontrollen mittels Real-Time PCR untersucht. In den meisten Fällen konnten keine signifikanten Unterschiede detektiert werden. Aber im Gegensatz zu den Ergebnissen der Subtraktionsbank, welche auf zwei Patienten beschränkt sind, zeigte der milde DMD Patient ein signifikant verringertes CSNK1A1-mRNA-Niveau im Vergleich zu den klassichen DMD Patienten ($P<0,05$). Ähnliche Ergebnisse ergaben sich für NFATc und JNK1 ($P<0,05$) im Vergleich zu Normalkontrollen. Desweiteren wurden für p21 und MYF5 im phänotypisch milden DMD Patienten das geringste Expressionsniveau bestimmt. Das UTRNA-mRNA-Niveau befand sich auf nahezu normalem Niveau im Vergleich zu DMD Patienten mit einer bis zu 14fachen Erhöhung. Diese Ergebnisse führten zur Entwicklung einer Therapiestrategie, die eine Proliferationserhöhung, um den dystrophen Prozess zu verlangsamen, mit einem gleichzeitig unveränderten oder gesteigerten UTRNA-Niveau vereint. Derzeitig konzentrieren sich derartige Therapiestrategien lediglich auf die Erhöhung von UTRNA.

Es folgten Untersuchungen, ob die p21 regulierte Myoblasten-Proliferation und das UTRNA Protein-Niveau verstärkt werden können, obwohl beide NFATc als Transkriptionsfaktor besitzen. Die gegensätzliche Wirkung von CSNK1A1 in DMD Patienten und die bemerkenswerte Rolle von Calcineurin innerhalb des Netzwerkes als CSNK1A1-Antagonist führten zu weiteren Untersuchungen. Diese beinhalteten die Modulation von beiden Proteinen und die Studie von deren Auswirkungen auf Proliferation und Genexpression von NFATc Zielgenen in humanen Skelettmuskelzellen (SkMCs) in Kultur. Unter Beachtung des Netzwerkmodells sollte eine Erhöhung der CSNK1A1 Expression das Niveau von UTRNA, MYF5 und p21 reduzieren, gefolgt von einer gesteigerten Proliferation der SkMCs. Die Genexpression von CSNK1A1 wurde durch die Transfektion von Plasmid-cDNA-Vektoren und siRNA erfolgreich verändert. Dennoch konnte kein eindeutiger Effekt auf Proliferation und Expression von UTRNA, MYF5 und p21 festgestellt werden. Calcineurin wurde durch Deflazacort und Cyclosporin A (CsA) beeinflusst. Die separate Verwendung von Deflazacort und CsA führte zu einem Anstieg der Proliferation ohne Einfluss auf die mRNA-Expression von MYF5 und p21 zu nehmen. Die Kombination von beiden, ähnlich praktiziert in einer klinischen Studie, ergab leicht erhöhte Proliferationswerte. Die Expression von UTRNA auf mRNA- und Protein-Ebene blieb durch CsA unverändert. Eine leichte Steigerung hingegen konnte durch Deflazacort festgestellt werden. Diese Wirkung wurde durch die Kombination noch verstärkt. Das mRNA-Niveau erhöhte sich auf das drei- bis siebenfache, während die

Proteinexpression eine Verdopplung aufzeigte. Dieser Effekt scheint sich auf DMD Patienten zu beschränken.

Zusammengefasst, weder die De- noch die Phosphorylierung von NFATc hat einen eindeutigen Einfluss auf das Expressionsniveau der Zielgene UTRNA, MYF5 und p21. Dies lässt vermuten, dass NFATc nicht der ausschlaggebende Transkriptionsfaktor für diese Zielgene ist und folglich auch nicht die p21 regulierte Zellproliferation beeinflusst. Dennoch konnte eine Erhöhung der Proliferation durch Deflazacort und/oder CsA induziert werden, was auf die Existenz weiterer Zielgene von NFATc oder auf andere Angriffspunkte von beiden Substanzen schließen lässt. Entgegen den Erwartungen aus dem Petri-Netz ist eine Steigerung der Proliferation ohne gleichzeitig die UTRNA Expression zu senken möglich. Diese Ergebnisse geben Anlass zur Entwicklung einer entsprechenden Therapiestrategie, zeigen aber auch die Notwendigkeit der Erweiterung des Petri-Netzes.

8 References

Ackermann,J. (2008). Mauritius maps. Unpublished analysis tool. Technical University of Applied Sciences Berlin.

Anderson,J.T., Rogers,R.P., and Jarrett,H.W. (1996). Ca2+-calmodulin binds to the carboxyl-terminal domain of dystrophin. J. Biol. Chem. *271*, 6605-6610.

Andersson,O., Kozlowski,M., Garachtchenko,T., Nikoloff,C., Lew,N., Litman,D.J., and Chaga,G. (2005). Determination of relative protein abundance by internally normalized ratio algorithm with antibody arrays. J. Proteome. Res. *4*, 758-767.

Angelini,C. (2007). The role of corticosteroids in muscular dystrophy: A critical appraisal. Muscle Nerve *36*, 424-435.

Angus,L.M., Chakkalakal,J.V., Mejat,A., Eibl,J.K., Belanger,G., Megeney,L.A., Chin,E.R., Schaeffer,L., Michel,R.N., and Jasmin,B.J. (2005). Calcineurin-NFAT signaling, together with GABP and peroxisome PGC-1{alpha}, drives utrophin gene expression at the neuromuscular junction. Am. J. Physiol Cell Physiol *289*, C908-C917.

Arbeitsausschuß Bioinformatik der DECHEMA e.V. (2006). Status- und Strategiepapier zur Systembiologie. DECHEMA *http://biotech.dechema.de/data/biotech_/ppsysbio.pdf*.

artsma-Rus,A., Fokkema,I., Verschuuren,J., Ginjaar,I., van,D.J., van Ommen,G.J., and den Dunnen,J.T. (2009). Theoretic applicability of antisense-mediated exon skipping for Duchenne muscular dystrophy mutations. Hum. Mutat. *in press*, 1-7.

artsma-Rus,A. and van Ommen,G.J. (2007). Antisense-mediated exon skipping: A versatile tool with therapeutic and research applications. RNA. *13*, 1609-1624.

Aurino,S. and Nigro,V. (2006). Readthrough strategies for stop codons in Duchenne muscular dystrophy. Acta Myol. *25*, 5-12.

Backhaus,K., Erichson,B., Plinke,W., and Weiber,R. (2000). Multivariate Analysemethoden. Eine Anwendungsorientierte Einfuehrung. (in German) Ed. 10, Berlin: Springer.

Bae,S.J., Oum,J.H., Sharma,S., Park,J., and Lee,S.W. (2002). In vitro selection of specific RNA inhibitors of NFATc. Biochem. Biophys. Res. Commun. *298*, 486-492.

Bakay,M., Zhao,P., Chen,J., and Hoffman,E.P. (2002). A web-accessible complete transcriptome of normal human and DMD muscle. Neuromuscul. Disord. *12 Suppl 1*, S125-S141.

Baker,P.E., Kearney,J.A., Gong,B., Merriam,A.P., Kuhn,D.E., Porter,J.D., and Rafael-Fortney,J.A. (2006). Analysis of gene expression differences between utrophin/dystrophin-deficient vs mdx skeletal muscles reveals a specific upregulation of slow muscle genes in limb muscles. Neurogenetics. *7*, 81-91.

Balghi,H., Sebille,S., Mondin,L., Cantereau,A., Constantin,B., Raymond,G., and Cognard,C. (2006). Mini-dystrophin expression down-regulates IP3-mediated calcium release events in resting dystrophin-deficient muscle cells. J. Gen. Physiol *128*, 219-230.

References

Basu,U., Gyrd-Hansen,M., Baby,S.M., Lozynska,O., Krag,T.O., Jensen,C.J., Frodin,M., and Khurana,T.S. (2007). Heregulin-induced epigenetic regulation of the utrophin-A promoter. FEBS Lett. *581*, 4153-4158.

Baumgarten,B. (1996). Petri-Netze - Grundlagen Und Anwendungen. 2nd Ed., Spektrum Akademischer Verlag, Heidelberg, Berlin, Oxford.

Baxevanis,A.D. and Ouellette,B.F.F. (2005). Bioinformatics - a Practical Guide to the Analysis of Genes and Proteins., London: John Wiley & Sons, Inc.

Beals,C.R., Sheridan,C.M., Turck,C.W., Gardner,P., and Crabtree,G.R. (1997). Nuclear export of NF-ATc enhanced by glycogen synthase kinase-3. Science *275*, 1930-1934.

Behrend,L., Milne,D.M., Stoter,M., Deppert,W., Campbell,L.E., Meek,D.W., and Knippschild,U. (2000). IC261, a specific inhibitor of the protein kinases casein kinase 1-delta and -epsilon, triggers the mitotic checkpoint and induces p53-dependent postmitotic effects. Oncogene *19*, 5303-5313.

Beroud,C. et al. (2007). Multiexon skipping leading to an artificial DMD protein lacking amino acids from exons 45 through 55 could rescue up to 63% of patients with Duchenne muscular dystrophy. Hum. Mutat. *28*, 196-202.

Bertoni,C. (2005). Oligonucleotide-mediated gene editing for neuromuscular disorders. Acta Myol. *24*, 194-201.

Bertoni,C. (2008). Clinical approaches in the treatment of Duchenne muscular dystrophy (DMD) using oligonucleotides. Front Biosci. *13*, 517-527.

Bialojan,C. and Takai,A. (1988). Inhibitory effect of a marine-sponge toxin, okadaic acid, on protein phosphatases. Specificity and kinetics. Biochem. J. *256*, 283-290.

Biggar,W.D., Harris,V.A., Eliasoph,L., and Alman,B. (2006). Long-term benefits of deflazacort treatment for boys with Duchenne muscular dystrophy in their second decade. Neuromuscul. Disord. *16*, 249-255.

Blau,H.M., Webster,C., Chiu,C.P., Guttman,S., and Chandler,F. (1983a). Differentiation properties of pure populations of human dystrophic muscle cells. Exp. Cell Res. *144*, 495-503.

Blau,H.M., Webster,C., and Pavlath,G.K. (1983b). Defective myoblasts identified in Duchenne muscular dystrophy. Proc. Natl. Acad. Sci. U. S. A *80*, 4856-4860.

Bode,A.M. and Dong,Z. (2007). The functional contrariety of JNK. Mol. Carcinog. *46*, 591-598.

Bogdanovich,S., Krag,T.O., Barton,E.R., Morris,L.D., Whittemore,L.A., Ahima,R.S., and Khurana,T.S. (2002). Functional improvement of dystrophic muscle by myostatin blockade. Nature *420*, 418-421.

Bogdanovich,S., Perkins,K.J., Krag,T.O., Whittemore,L.A., and Khurana,T.S. (2005). Myostatin propeptide-mediated amelioration of dystrophic pathophysiology. FASEB J. *19*, 543-549.

REFERENCES

Bonuccelli,G., Sotgia,F., Capozza,F., Gazzerro,E., Minetti,C., and Lisanti,M.P. (2007). Localized treatment with a novel FDA-approved proteasome inhibitor blocks the degradation of dystrophin and dystrophin-associated proteins in mdx mice. Cell Cycle *6*, 1242-1248.

Bradley,L., Yaworsky,P.J., and Walsh,F.S. (2008). Myostatin as a therapeutic target for musculoskeletal disease. Cell Mol. Life Sci. *65*, 2119-2124.

Briguet,A. *et al.* (2008). Effect of calpain and proteasome inhibition on Ca2+-dependent proteolysis and muscle histopathology in the mdx mouse. FASEB J. *22*, 4190-4200.

Buyse,G.M. *et al.* (2009). Long-term blinded placebo-controlled study of SNT-MC17/idebenone in the dystrophin deficient mdx mouse: cardiac protection and improved exercise performance. Eur. Heart J. *30*, 116-124.

Calderilla-Barbosa,L., Ortega,A., and Cisneros,B. (2006). Phosphorylation of dystrophin Dp71d by Ca2+/calmodulin-dependent protein kinase II modulates the Dp71d nuclear localization in PC12 cells. J. Neurochem. *98*, 713-722.

Campeau,P., Chapdelaine,P., Seigneurin-Venin,S., Massie,B., and Tremblay,J.P. (2001). Transfection of large plasmids in primary human myoblasts. Gene Ther. *8*, 1387-1394.

Cascante,M., Boros,L.G., Comin-Anduix,B., de,A.P., Centelles,J.J., and Lee,P.W. (2002). Metabolic control analysis in drug discovery and disease. Nat. Biotechnol. *20*, 243-249.

Chaga,G.S. (2008). Antibody arrays for determination of relative protein abundances. Methods Mol. Biol. *441*, 129-151.

Chakkalakal,J.V., Harrison,M.A., Carbonetto,S., Chin,E., Michel,R.N., and Jasmin,B.J. (2004). Stimulation of calcineurin signaling attenuates the dystrophic pathology in mdx mice. Hum. Mol. Genet. *13*, 379-388.

Chakkalakal,J.V., Michel,S.A., Chin,E.R., Michel,R.N., and Jasmin,B.J. (2006). Targeted inhibition of Ca2+ /calmodulin signaling exacerbates the dystrophic phenotype in mdx mouse muscle. Hum. Mol. Genet. *15*, 1423-1435.

Chakkalakal,J.V., Stocksley,M.A., Harrison,M.A., Angus,L.M., schenes-Furry,J., St-Pierre,S., Megeney,L.A., Chin,E.R., Michel,R.N., and Jasmin,B.J. (2003). Expression of utrophin A mRNA correlates with the oxidative capacity of skeletal muscle fiber types and is regulated by calcineurin/NFAT signaling. Proc. Natl. Acad. Sci. U. S. A *100*, 7791-7796.

Chaouiya,C. (2007). Petri net modelling of biological networks. Brief. Bioinform. *8*, 210-219.

Chaouiya,C., Remy,E., Ruet,P., and et al. (2004). Qualitative modelling of genetic networks: from logical regulatory graphs to standard Petri nets. LectNotes Comp Sci *3099*, 137-56.

Chen,M. and Hofestadt,R. (2006). A medical bioinformatics approach for metabolic disorders: biomedical data prediction, modeling, and systematic analysis. J. Biomed. Inform. *39*, 147-159.

Chin,E.R., Olson,E.N., Richardson,J.A., Yang,Q., Humphries,C., Shelton,J.M., Wu,H., Zhu,W., Bassel-Duby,R., and Williams,R.S. (1998). A calcineurin-dependent transcriptional pathway controls skeletal muscle fiber type. Genes Dev. *12*, 2499-2509.

REFERENCES

Cho,J.S., Lee,Y.J., Shin,K.S., Jeong,S., Park,J., and Lee,S.W. (2004). In vitro selection of specific RNA aptamers for the NFAT DNA binding domain. Mol. Cells *18*, 17-23.

Chow,C.W., Dong,C., Flavell,R.A., and Davis,R.J. (2000). c-Jun NH(2)-terminal kinase inhibits targeting of the protein phosphatase calcineurin to NFATc1. Mol. Cell Biol. *20*, 5227-5234.

Constantin,B., Sebille,S., and Cognard,C. (2006). New insights in the regulation of calcium transfers by muscle dystrophin-based cytoskeleton: implications in DMD. J. Muscle Res. Cell Motil. *27*, 375-386.

Crabtree,G.R. and Olson,E.N. (2002). NFAT signaling: choreographing the social lives of cells. Cell *109 Suppl:S67-79.*, S67-S79.

De,L.A. *et al.* (2005). A multidisciplinary evaluation of the effectiveness of cyclosporine a in dystrophic mdx mice. Am. J. Pathol. *166*, 477-489.

Desantis,A., Onori,A., Di Certo,M.G., Mattei,E., Fanciulli,M., Passananti,C., and Corbi,N. (2009). Novel activation domain derived from Che-1 cofactor coupled with the artificial protein Jazz drives utrophin upregulation. Neuromuscul. Disord. *19*, 158-162.

Dickson,G., Gower,H.J., Barton,C.H., Prentice,H.M., Elsom,V.L., Moore,S.E., Cox,R.D., Quinn,C., Putt,W., and Walsh,F.S. (1987). Human muscle neural cell adhesion molecule (N-CAM): identification of a muscle-specific sequence in the extracellular domain. Cell *50*, 1119-1130.

Dogra,C., Srivastava,D.S., and Kumar,A. (2008). Protein-DNA array-based identification of transcription factor activities differentially regulated in skeletal muscle of normal and dystrophin-deficient mdx mice. Mol. Cell Biochem. *312*, 17-24.

Dong,C., Yang,D.D., Wysk,M., Whitmarsh,A.J., Davis,R.J., and Flavell,R.A. (1998). Defective T cell differentiation in the absence of Jnk1. Science *282*, 2092-2095.

Doran,P., Wilton,S.D., Fletcher,S., and Ohlendieck,K. (2009). Proteomic profiling of antisense-induced exon skipping reveals reversal of pathobiochemical abnormalities in dystrophic mdx diaphragm. Proteomics. *9*, 671-685.

Dounay,A.B. and Forsyth,C.J. (2002). Okadaic acid: the archetypal serine/threonine protein phosphatase inhibitor. Curr. Med. Chem. *9*, 1939-1980.

Duan,D. (2008). Myodys, a full-length dystrophin plasmid vector for Duchenne and Becker muscular dystrophy gene therapy. Curr. Opin. Mol. Ther. *10*, 86-94.

Dudley,R.W., Lu,Y., Gilbert,R., Matecki,S., Nalbantoglu,J., Petrof,B.J., and Karpati,G. (2004). Sustained improvement of muscle function one year after full-length dystrophin gene transfer into mdx mice by a gutted helper-dependent adenoviral vector. Hum. Gene Ther. *15*, 145-156.

Dunn,S.E., Simard,A.R., Prud'homme,R.A., and Michel,R.N. (2002). Calcineurin and skeletal muscle growth. Nat. Cell Biol. *4*, E46-E47.

Ellis,R.J. and Minton,A.P. (2003). Cell biology: join the crowd. Nature *425*, 27-28.

Endesfelder,S. (2004). Untersuchungen zur Regulation von Proliferation und Differenzierung von Dystrophin-defizienten Muskelzellen mittels p21 antisense Oligonukleotiden: eine neue Strategie zur Therapie der Muskeldystrophie Duchenne. (in German). Dissertation. Humboldt Universitaet Berlin. Mathematisch-Naturwissenschaftliche Fakultät I.

Endesfelder,S., Bucher,S., Kliche,A., Reszka,R., and Speer,A. (2003). Transfection of normal primary human skeletal myoblasts with p21 and p57 antisense oligonucleotides to improve their proliferation: a first step towards an alternative molecular therapy approach of Duchenne muscular dystrophy. J. Mol. Med. *81*, 355-362.

Endesfelder,S., Kliche,A., Lochmuller,H., von,M.A., and Speer,A. (2005). Antisense oligonucleotides and short interfering RNAs silencing the cyclin-dependent kinase inhibitor p21 improve proliferation of Duchenne muscular dystrophy patients' primary skeletal myoblasts. J. Mol. Med. *83*, 64-71.

Endesfelder,S., Krahn,A., Kreuzer,K.A., Lass,U., Schmidt,C.A., Jahrmarkt,C., von,M.A., and Speer,A. (2000). Elevated p21 mRNA level in skeletal muscle of DMD patients and mdx mice indicates either an exhausted satellite cell pool or a higher p21 expression in dystrophin-deficient cells per se. J. Mol. Med. *78*, 569-574.

Ervasti,J.M. (2007). Dystrophin, its interactions with other proteins, and implications for muscular dystrophy. Biochim. Biophys. Acta *1772*, 108-117.

Ervasti,J.M. and Sonnemann,K.J. (2008). Biology of the striated muscle dystrophin-glycoprotein complex. Int. Rev. Cytol. *265*, 191-225.

Espinos,E., Liu,J.H., Bader,C.R., and Bernheim,L. (2001). Efficient non-viral DNA-mediated gene transfer to human primary myoblasts using electroporation. Neuromuscul. Disord. *11*, 341-349.

Evellin,S., Nolte,J., Tysack,K., Vom,D.F., Thiel,M., Weernink,P.A., Jakobs,K.H., Webb,E.J., Lomasney,J.W., and Schmidt,M. (2002). Stimulation of phospholipase C-epsilon by the M3 muscarinic acetylcholine receptor mediated by cyclic AMP and the GTPase Rap2B. J. Biol. Chem. *277*, 16805-16813.

Fardeau,M., Braun,S., Romero,N.B., Hogrel,J.Y., Rouche,A., Ortega,V., Mourot,B., Squiban,P., Benveniste,O., and Herson,S. (2005). [About a phase I gene therapy clinical trial with a full-length dystrophin gene-plasmid in Duchenne/Becker muscular dystrophy]. J. Soc. Biol. *199*, 29-32.

Ferreiro,V., Giliberto,F., Muniz,G.M., Francipane,L., Marzese,D.M., Mampel,A., Roque,M., Frechtel,G.D., and Szijan,I. (2009). Asymptomatic Becker muscular dystrophy in a family with a multiexon deletion. Muscle Nerve. *39*, 239-243.

Fieber,M. (2004). Design and implementation of a generic and adaptive tool for graph manipulation. (in German). Master thesis. Brandenburg University of Technology Cottbus. Computer Science Dept.

Fisher,R., Tinsley,J.M., Phelps,S.R., Squire,S.E., Townsend,E.R., Martin,J.E., and Davies,K.E. (2001). Non-toxic ubiquitous over-expression of utrophin in the mdx mouse. Neuromuscul. Disord. *11*, 713-721.

Friday,B.B. and Pavlath,G.K. (2001). A calcineurin- and NFAT-dependent pathway regulates Myf5 gene expression in skeletal muscle reserve cells. J. Cell Sci. *114*, 303-310.

Grafahrend-Belau,E. (2006). Classification of T-invariants in biochemical Petri nets based on different cluster analysis techniques (in German). Master thesis. Technical University of Applied Sciences Berlin.

Grafahrend-Belau,E., Schreiber,F., Heiner,M., Sackmann,A., Junker,B.H., Grunwald,S., Speer,A., Winder,K., and Koch,I. (2008a). Modularization of biochemical networks based on classification of Petri net t-invariants. BMC. Bioinformatics. *9*, 90.

Grafahrend-Belau,E., Weise,S., Koschutzki,D., Scholz,U., Junker,B.H., and Schreiber,F. (2008b). MetaCrop: a detailed database of crop plant metabolism. Nucleic Acids Res. *36*, D954-D958.

Groenendyk,J., Lynch,J., and Michalak,M. (2004). Calreticulin, Ca2+, and calcineurin - signaling from the endoplasmic reticulum. Mol. Cells *17*, 383-389.

Grounds,M.D. and Davies,K.E. (2007). The allure of stem cell therapy for muscular dystrophy. Neuromuscul. Disord. *17*, 206-208.

Grunwald,S. and Speer,A. (2007). Efficient transfection of primary human skeletal myoblasts using FuGENE® HD transfection reagent. Roche Biochemica *3*, 26-27.

Grunwald,S., Speer,A., Ackermann,J., and Koch,I. (2008). Petri net modelling of gene regulation of the Duchenne muscular dystrophy. Biosystems *92*, 189-205.

Gyrd-Hansen,M., Krag,T.O., Rosmarin,A.G., and Khurana,T.S. (2002). Sp1 and the ets-related transcription factor complex GABP alpha/beta functionally cooperate to activate the utrophin promoter. J. Neurol. Sci. *197*, 27-35.

Hacein-Bey-Abina,S. *et al.* (2003a). A serious adverse event after successful gene therapy for X-linked severe combined immunodeficiency. N. Engl. J. Med. *348*, 255-256.

Hacein-Bey-Abina,S. *et al.* (2003b). LMO2-associated clonal T cell proliferation in two patients after gene therapy for SCID-X1. Science *302*, 415-419.

Harbour,J.W., Luo,R.X., Dei,S.A., Postigo,A.A., and Dean,D.C. (1999). Cdk phosphorylation triggers sequential intramolecular interactions that progressively block Rb functions as cells move through G1. Cell *98*, 859-869.

Haslett,J.N., Sanoudou,D., Kho,A.T., Bennett,R.R., Greenberg,S.A., Kohane,I.S., Beggs,A.H., and Kunkel,L.M. (2002). Gene expression comparison of biopsies from Duchenne muscular dystrophy (DMD) and normal skeletal muscle. Proc. Natl. Acad. Sci. U. S. A *99*, 15000-15005.

Haslett,J.N., Sanoudou,D., Kho,A.T., Han,M., Bennett,R.R., Kohane,I.S., Beggs,A.H., and Kunkel,L.M. (2003). Gene expression profiling of Duchenne muscular dystrophy skeletal muscle. Neurogenetics. *4*, 163-171.

Hattori,N., Kaido,M., Nishigaki,T., Inui,K., Fujimura,H., Nishimura,T., Naka,T., and Hazama,T. (1999). Undetectable dystrophin can still result in a relatively benign phenotype of dystrophinopathy. Neuromuscul. Disord. *9*, 220-226.

Heemskerk,H.A., De Winter,C.L., de Kimpe,S.J., van Kuik-Romeijn,P., Heuvelmans,N., Platenburg,G.J., van Ommen,G.J., van Deutekom,J.C., and artsma-Rus,A. (2009). In vivo comparison of 2'-O-methyl phosphorothioate and morpholino antisense oligonucleotides for Duchenne muscular dystrophy exon skipping. J. Gene Med. *in press*.

Heiner,M. and Koch,I. (2004). Petri net based model validation in systems biology. Proceedings of the 25th International Conference on Applications and Theory of Petri Nets, Bologna. *LCNS 3099*, 216-37.

Heiner,M., Koch,I., and Will,J. (2004). Model validation of biological pathways using Petri nets--demonstrated for apoptosis. Biosystems *75*, 15-28.

Heinrich,R. and Schuster,S. (1998). The modelling of metabolic systems. Structure, control and optimality. Biosystems *47*, 61-77.

Hinkes,B. *et al.* (2006). Positional cloning uncovers mutations in PLCE1 responsible for a nephrotic syndrome variant that may be reversible. Nat. Genet *38*, 1397-1405.

Hirawat,S. *et al.* (2007). Safety, tolerability, and pharmacokinetics of PTC124, a nonaminoglycoside nonsense mutation suppressor, following single- and multiple-dose administration to healthy male and female adult volunteers. J. Clin. Pharmacol. *47*, 430-444.

Hoffman,E.P., Knudson,C.M., Campbell,K.P., and Kunkel,L.M. (1987). Subcellular fractionation of dystrophin to the triads of skeletal muscle. Nature *330*, 754-758.

Hogan,P.G., Chen,L., Nardone,J., and Rao,A. (2003). Transcriptional regulation by calcium, calcineurin, and NFAT. Genes Dev. *17*, 2205-2232.

Hood,L., Heath,J.R., Phelps,M.E., and Lin,B. (2004). Systems biology and new technologies enable predictive and preventative medicine. Science *306*, 640-643.

Hopf,F.W., Turner,P.R., and Steinhardt,R.A. (2007). Calcium misregulation and the pathogenesis of muscular dystrophy. Subcell. Biochem. *45*, 429-464.

Houde,S., Filiatrault,M., Fournier,A., Dube,J., D'Arcy,S., Berube,D., Brousseau,Y., Lapierre,G., and Vanasse,M. (2008). Deflazacort use in duchenne muscular dystrophy: an 8-year follow-up. Pediatr. Neurol *38*, 200-206.

Im,S.H. and Rao,A. (2004). Activation and deactivation of gene expression by Ca^{2+}/calcineurin-NFAT-mediated signaling. Mol. Cells *18*, 1-9.

Ishida,H. and Vogel,H.J. (2006). Protein-peptide interaction studies demonstrate the versatility of calmodulin target protein binding. Protein Pept. Lett. *13*, 455-465.

Kacser,H. (1995). Recent developments beyond metabolic control analysis. Biochem. Soc. Trans. *23*, 387-391.

Kambadur,R., Sharma,M., Smith,T.P., and Bass,J.J. (1997). Mutations in myostatin (GDF8) in double-muscled Belgian Blue and Piedmontese cattle. Genome Res. *7*, 910-916.

Keiper,M. *et al.* (2004). Epac- and Ca^{2+} -controlled activation of Ras and extracellular signal-regulated kinases by Gs-coupled receptors. J. Biol. Chem. *279*, 46497-46508.

Knippschild,U., Gocht,A., Wolff,S., Huber,N., Lohler,J., and Stoter,M. (2005a). The casein kinase 1 family: participation in multiple cellular processes in eukaryotes. Cell Signal. *17*, 675-689.

Knippschild,U., Wolff,S., Giamas,G., Brockschmidt,C., Wittau,M., Wurl,P.U., Eismann,T., and Stoter,M. (2005b). The role of the casein kinase 1 (CK1) family in different signaling pathways linked to cancer development. Onkologie. *28*, 508-514.

Koch,I. and Heiner,M. (2008). Petri Nets in Biological Networkanalysis. In: Analysis of biological networks, ed. F.Schreiber and B.Junker Wiley & Sons, 139-179.

Koch,I., Junker,B.H., and Heiner,M. (2005). Application of Petri net theory for modelling and validation of the sucrose breakdown pathway in the potato tuber. Bioinformatics. *21*, 1219-1226.

Koenig,M., Monaco,A.P., and Kunkel,L.M. (1988). The complete sequence of dystrophin predicts a rod-shaped cytoskeletal protein. Cell *53*, 219-228.

Kolodziejczyk,S.M., Walsh,G.S., Balazsi,K., Seale,P., Sandoz,J., Hierlihy,A.M., Rudnicki,M.A., Chamberlain,J.S., Miller,F.D., and Megeney,L.A. (2001). Activation of JNK1 contributes to dystrophic muscle pathogenesis. Curr. Biol. *11*, 1278-1282.

Korinthenberg,R. (2008). Phase II - III: Immunsuppressive Therapie bei Muskeldystrophie Duchenne. Study UKF000560. Kinderklinik der Uniklinik Freiburg. *http://www.zks.uni-freiburg.de/uklreg/php/show_study.php?STUDIEN_ID=000560&kindOfSearch=frei&lang=DE.*.

Kosek,D.J., Kim,J.S., Petrella,J.K., Cross,J.M., and Bamman,M.M. (2006). Efficacy of 3 days/wk resistance training on myofiber hypertrophy and myogenic mechanisms in young vs. older adults. J. Appl. Physiol *101*, 531-544.

Laemmli,U.K. (1970). Cleavage of structural proteins during the assembly of the head of bacteriophage T4. Nature *227*, 680-685.

Laure,L., Suel,L., Roudaut,C., Bourg,N., Ouali,A., Bartoli,M., Richard,I., and Daniele,N. (2009). Cardiac ankyrin repeat protein is a marker of skeletal muscle pathological remodelling. FEBS J. *276*, 669-684.

Lautenbach,K. (1973). Exact liveness conditions of a Petri Net class (in German). GMD Report 82, Bonn.

Lee,D.Y., Zimmer,R., Lee,S.Y., and Park,S. (2006). Colored Petri net modeling and simulation of signal transduction pathways. Metab Eng *8*, 112-122.

Legardinier,S., Legrand,B., Raguenes-Nicol,C., Bondon,A., Hardy,S., Tascon,C., Le,R.E., and Hubert,J.F. (2009). A two-amino-acid mutation encountered in a Duchenne muscular dystrophy decreases stability of the R23 spectrin-like repeat of dystrophin. J. Biol. Chem. *in press*.

Legewie,S., Bluthgen,N., and Herzel,H. (2006). Mathematical modeling identifies inhibitors of apoptosis as mediators of positive feedback and bistability. PLoS. Comput. Biol. *2*, e120.

Li,D., Long,C., Yue,Y., and Duan,D. (2009). Sub-physiological sarcoglycan expression contributes to compensatory muscle protection in mdx mice. Hum. Mol. Genet. *in press*.

Link,H. and Weuster-Botz,D. (2007). Steady-state analysis of metabolic pathways: comparing the double modulation method and the lin-log approach. Metab Eng *9*, 433-441.

Love,D.R., Hill,D.F., Dickson,G., Spurr,N.K., Byth,B.C., Marsden,R.F., Walsh,F.S., Edwards,Y.H., and Davies,K.E. (1989). An autosomal transcript in skeletal muscle with homology to dystrophin. Nature *339*, 55-58.

Lu,S., Jenster,G., and Epner,D.E. (2000). Androgen induction of cyclin-dependent kinase inhibitor p21 gene: role of androgen receptor and transcription factor Sp1 complex. Mol. Endocrinol. *14*, 753-760.

Macian,F. (2005). NFAT proteins: key regulators of T-cell development and function. Nat. Rev. Immunol. *5*, 472-484.

Madhavan,R. and Jarrett,H.W. (1994). Calmodulin-activated phosphorylation of dystrophin. Biochemistry *33*, 5797-5804.

Madhavan,R. and Jarrett,H.W. (1999). Phosphorylation of dystrophin and alpha-syntrophin by Ca(2+)-calmodulin dependent protein kinase II. Biochim. Biophys. Acta *1434*, 260-274.

Manceau,M., Gros,J., Savage,K., Thome,V., McPherron,A., Paterson,B., and Marcelle,C. (2008). Myostatin promotes the terminal differentiation of embryonic muscle progenitors. Genes Dev. *22*, 668-681.

Manzur,A., Kuntzer,T., Pike,M., and Swan,A. (2008a). Glucocorticoid corticosteroids for Duchenne muscular dystrophy. Cochrane. Database. Syst. Rev. CD003725.

Manzur,A.Y., Kinali,M., and Muntoni,F. (2008b). Update on the management of Duchenne muscular dystrophy. Arch. Dis. Child. *93*, 986-990.

Marwan,W., Sujatha,A., and Starostzik,C. (2005). Reconstructing the regulatory network controlling commitment and sporulation in Physarum polycephalum based on hierarchical Petri Net modelling and simulation. J. Theor. Biol. *236*, 349-365.

Matsumura,C.Y., Pertille,A., Albuquerque,T.C., Santo,N.H., and Marques,M.J. (2009). Diltiazem and verapamil protect dystrophin-deficient muscle fibers of MDX mice from degeneration: A potential role in calcium buffering and sarcolemmal stability. Muscle Nerve. *39*, 167-176.

Matsuno,H., Doi,A., Nagasaki,M., and Miyano,S. (2000). Hybrid Petri net representation of gene regulatory network. Pac. Symp. Biocomput. 341-352.

Matsuno,H., Tanaka,Y., Aoshima,H., Doi,A., Matsui,M., and Miyano,S. (2003). Biopathways representation and simulation on hybrid functional Petri net. In Silico. Biol. *3*, 389-404.

Mattei,E. *et al.* (2007). Utrophin up-regulation by an artificial transcription factor in transgenic mice. PLoS. ONE. *2*, e774.

Mavrogeni,S., Papavasiliou,A., Douskou,M., Kolovou,G., Papadopoulou,E., and Cokkinos,D.V. (2009). Effect of deflazacort on cardiac and sternocleidomastoid muscles in

Duchenne muscular dystrophy: A magnetic resonance imaging study. Eur. J. Paediatr. Neurol. *13*, 34-40.

Mesaeli,N., Nakamura,K., Zvaritch,E., Dickie,P., Dziak,E., Krause,K.H., Opas,M., MacLennan,D.H., and Michalak,M. (1999). Calreticulin is essential for cardiac development. J. Cell Biol. *144*, 857-868.

Michel,R.N., Chin,E.R., Chakkalakal,J.V., Eibl,J.K., and Jasmin,B.J. (2007). Ca2+/calmodulin-based signalling in the regulation of the muscle fibre phenotype and its therapeutic potential via modulation of utrophin A and myostatin expression. Appl. Physiol Nutr. Metab *32*, 921-929.

Michel,R.N., Dunn,S.E., and Chin,E.R. (2004). Calcineurin and skeletal muscle growth. Proc. Nutr. Soc. *63*, 341-349.

Mitrpant,C., Fletcher,S., Iversen,P.L., and Wilton,S.D. (2009). By-passing the nonsense mutation in the 4 CV mouse model of muscular dystrophy by induced exon skipping. J. Gene Med. *11*, 46-56.

Miura,P. and Jasmin,B.J. (2006). Utrophin upregulation for treating Duchenne or Becker muscular dystrophy: how close are we? Trends Mol. Med. *12*, 122-129.

Molkentin,J.D., Lu,J.R., Antos,C.L., Markham,B., Richardson,J., Robbins,J., Grant,S.R., and Olson,E.N. (1998). A calcineurin-dependent transcriptional pathway for cardiac hypertrophy. Cell *93*, 215-228.

Monaco,A.P., Neve,R.L., Colletti-Feener,C., Bertelson,C.J., Kurnit,D.M., and Kunkel,L.M. (1986). Isolation of candidate cDNAs for portions of the Duchenne muscular dystrophy gene. Nature *323*, 646-650.

Morton,S., Davis,R.J., McLaren,A., and Cohen,P. (2003). A reinvestigation of the multisite phosphorylation of the transcription factor c-Jun. EMBO J. *22*, 3876-3886.

Muntoni,F., Bushby,K.D., and van,O.G. (2008). 149th ENMC International Workshop and 1st TREAT-NMD Workshop on: "Planning Phase I/II Clinical trials using Systemically Delivered Antisense Oligonucleotides in Duchenne Muscular Dystrophy". Neuromuscul. Disord. *18*, 268-275.

Muntoni,F., Torelli,S., and Ferlini,A. (2003). Dystrophin and mutations: one gene, several proteins, multiple phenotypes. Lancet Neurol. *2*, 731-740.

Muntoni,F. and Wells,D. (2007). Genetic treatments in muscular dystrophies. Curr. Opin. Neurol. *20*, 590-594.

Nakamura,A., Yoshida,K., Ueda,H., Takeda,S., and Ikeda,S. (2005). Up-regulation of mitogen activated protein kinases in mdx skeletal muscle following chronic treadmill exercise. Biochim. Biophys. Acta *1740*, 326-331.

Nakatani,M. *et al.* (2008). Transgenic expression of a myostatin inhibitor derived from follistatin increases skeletal muscle mass and ameliorates dystrophic pathology in mdx mice. FASEB J. *22*, 477-487.

Neri,M. *et al.* (2007). Dystrophin levels as low as 30% are sufficient to avoid muscular dystrophy in the human. Neuromuscul. Disord. *17*, 913-918.

Neuhaus,P., Oustanina,S., Loch,T., Kruger,M., Bober,E., Dono,R., Zeller,R., and Braun,T. (2003). Reduced mobility of fibroblast growth factor (FGF)-deficient myoblasts might contribute to dystrophic changes in the musculature of FGF2/FGF6/mdx triple-mutant mice. Mol Cell Biol. *23*, 6037-6048.

Noguchi,S., Tsukahara,T., Fujita,M., Kurokawa,R., Tachikawa,M., Toda,T., Tsujimoto,A., Arahata,K., and Nishino,I. (2003). cDNA microarray analysis of individual Duchenne muscular dystrophy patients. Hum Mol Genet *12*, 595-600.

Oak,S.A., Zhou,Y.W., and Jarrett,H.W. (2003). Skeletal muscle signaling pathway through the dystrophin glycoprotein complex and Rac1. J. Biol. Chem. *278*, 39287-39295.

Odom,G.L., Gregorevic,P., and Chamberlain,J.S. (2007). Viral-mediated gene therapy for the muscular dystrophies: successes, limitations and recent advances. Biochim. Biophys. Acta *1772*, 243-262.

Ohshima,S., Shin,J.H., Yuasa,K., Nishiyama,A., Kira,J., Okada,T., and Takeda,S. (2009). Transduction efficiency and immune response associated with the administration of AAV8 vector into dog skeletal muscle. Mol. Ther. *17*, 73-80.

Onori,A., Desantis,A., Buontempo,S., Di Certo,M.G., Fanciulli,M., Salvatori,L., Passananti,C., and Corbi,N. (2007). The artificial 4-zinc-finger protein Bagly binds human utrophin promoter A at the endogenous chromosomal site and activates transcription. Biochem. Cell Biol. *85*, 358-365.

Pampinella,F., Lechardeur,D., Zanetti,E., MacLachlan,I., Benharouga,M., Lukacs,G.L., and Vitiello,L. (2002). Analysis of differential lipofection efficiency in primary and established myoblasts. Mol. Ther. *5*, 161-169.

Perkins,K.J. and Davies,K.E. (2003). Ets, Ap-1 and GATA factor families regulate the utrophin B promoter: potential regulatory mechanisms for endothelial-specific expression. FEBS Lett. *538*, 168-172.

Pescatori,M. *et al.* (2007). Gene expression profiling in the early phases of DMD: a constant molecular signature characterizes DMD muscle from early postnatal life throughout disease progression. FASEB J. *21*, 1210-1226.

Peter,A.K., Ko,C.Y., Kim,M.H., Hsu,N., Ouchi,N., Rhie,S., Izumiya,Y., Zeng,L., Walsh,K., and Crosbie,R.H. (2009). Myogenic Akt signaling upregulates the utrophin-glycoprotein complex and promotes sarcolemma stability in muscular dystrophy. Hum. Mol. Genet. *18*, 318-327.

Phillips,M.F. and Quinlivan,R. (2008). Calcium antagonists for Duchenne muscular dystrophy. Cochrane. Database. Syst. Rev. CD004571.

Philpott,N.J. and Thrasher,A.J. (2007). Use of nonintegrating lentiviral vectors for gene therapy. Hum Gene Ther. *18*, 483-489.

Politano,L., Nigro,G., Nigro,V., Piluso,G., Papparella,S., Paciello,O., and Comi,L.I. (2003). Gentamicin administration in Duchenne patients with premature stop codon. Preliminary results. Acta Myol. *22*, 15-21.

Pongratz,D.E., Reimers,C.D., Hahn,D., Naegele,M., and Mueller-Felber,W. (1990). Atlas Der Muskelkrankheiten., Muenchen-Wien-Baltimore: Urban & Schwarzenberg Bei Elsevier.

Pradhan,S. (2004). Valley sign in Becker muscular dystrophy and outliers of Duchenne and Becker muscular dystrophy. Neurol. India *52*, 203-205.

Prior,T.W., Bartolo,C., Papp,A.C., Snyder,P.J., Sedra,M.S., Burghes,A.H., Kissel,J.T., Luquette,M.H., Tsao,C.Y., and Mendell,J.R. (1997). Dystrophin expression in a Duchenne muscular dystrophy patient with a frame shift deletion. Neurology *48*, 486-488.

Qiao,C., Li,J., Jiang,J., Zhu,X., Wang,B., Li,J., and Xiao,X. (2008). Myostatin Propeptide Gene Delivery by Adeno-Associated Virus Serotype 8 Vectors Enhances Muscle Growth and Ameliorates Dystrophic Phenotypes in mdx Mice. Hum. Gene Ther. *19*, 241-254.

Quenneville,S.P., Chapdelaine,P., Rousseau,J., Beaulieu,J., Caron,N.J., Skuk,D., Mills,P., Olivares,E.C., Calos,M.P., and Tremblay,J.P. (2004). Nucleofection of muscle-derived stem cells and myoblasts with phiC31 integrase: stable expression of a full-length-dystrophin fusion gene by human myoblasts. Mol. Ther. *10*, 679-687.

Rando,T.A. (2007). Non-viral gene therapy for Duchenne muscular dystrophy: progress and challenges. Biochim. Biophys. Acta *1772*, 263-271.

Reddy,V.N., Mavrovouniotis,M.L., and Liebman,M.N. (1993). Petri net representations in metabolic pathways. Proc. Int. Conf. Intell. Syst. Mol. Biol. *1*, 328-336.

Roberts,R.G. (2001). Dystrophins and dystrobrevins. Genome Biol. *2*, REVIEWS3006.

Roberts,R.G., Gardner,R.J., and Bobrow,M. (1994). Searching for the 1 in 2,400,000: a review of dystrophin gene point mutations. Hum. Mutat. *4*, 1-11.

Rodova,M., Brownback,K., and Werle,M.J. (2004). Okadaic acid augments utrophin in myogenic cells. Neurosci. Lett. *363*, 163-167.

Romano,G. (2005). Current development of lentiviral-mediated gene transfer. Drug News Perspect. *18*, 128-134.

Sackmann,A., Formanowicz,D., Formanowicz,P., Koch,I., and Blazewicz,J. (2007). An analysis of the Petri net based model of the human body iron homeostasis process. Comput. Biol. Chem. *31*, 1-10.

Sackmann,A., Heiner,M., and Koch,I. (2006). Application of Petri net based analysis techniques to signal transduction pathways. BMC. Bioinformatics. *7*, 482.

Saito,A., Nagasaki,M., Doi,A., Ueno,K., and Miyano,S. (2006). Cell fate simulation model of gustatory neurons with MicroRNAs double-negative feedback loop by hybrid functional Petri net with extension. Genome Inform. *17*, 100-111.

Sakamoto,M., Yuasa,K., Yoshimura,M., Yokota,T., Ikemoto,T., Suzuki,M., Dickson,G., Miyagoe-Suzuki,Y., and Takeda,S. (2002). Micro-dystrophin cDNA ameliorates dystrophic phenotypes when introduced into mdx mice as a transgene. Biochem. Biophys. Res. Commun. *293*, 1265-1272.

Salter,M., Knowles,R.G., and Pogson,C.I. (1994). Metabolic control. Essays Biochem. *28*, 1-12.

Santini,M.P., Talora,C., Seki,T., Bolgan,L., and Dotto,G.P. (2001). Cross talk among calcineurin, Sp1/Sp3, and NFAT in control of p21(WAF1/CIP1) expression in keratinocyte differentiation. Proc. Natl. Acad. Sci. U. S. A *98*, 9575-9580.

Schertzer,J.D., van der,P.C., Shavlakadze,T., Grounds,M.D., and Lynch,G.S. (2008). Muscle-specific overexpression of IGF-I improves E-C coupling in skeletal muscle fibers from dystrophic mdx mice. Am. J. Physiol Cell Physiol *294*, C161-C168.

Schindelhauer,D. and Laner,A. (2002). Visible transient expression of EGFP requires intranuclear injection of large copy numbers. Gene Ther. *9*, 727-730.

Schmidt,M., Evellin,S., Weernink,P.A., von,D.F., Rehmann,H., Lomasney,J.W., and Jakobs,K.H. (2001). A new phospholipase-C-calcium signalling pathway mediated by cyclic AMP and a Rap GTPase. Nat. Cell Biol. *3*, 1020-1024.

Schmitz,A. and Famulok,M. (2007). Chemical biology: ignore the nonsense. Nature *447*, 42-43.

Schreiber,A., Smith,W.L., Ionasescu,V., Zellweger,H., Franken,E.A., Dunn,V., and Ehrhardt,J. (1987). Magnetic resonance imaging of children with Duchenne muscular dystrophy. Pediatr. Radiol. *17*, 495-497.

Schroder,A.R., Shinn,P., Chen,H., Berry,C., Ecker,J.R., and Bushman,F. (2002). HIV-1 integration in the human genome favors active genes and local hotspots. Cell *110*, 521-529.

Schulz,R.A. and Yutzey,K.E. (2004). Calcineurin signaling and NFAT activation in cardiovascular and skeletal muscle development. Dev. Biol. *266*, 1-16.

Schuster,S., Dandekar,T., and Fell,D.A. (1999). Detection of elementary flux modes in biochemical networks: a promising tool for pathway analysis and metabolic engineering. Trends Biotechnol. *17*, 53-60.

Scime,A. and Rudnicki,M.A. (2008). Molecular-targeted therapy for Duchenne muscular dystrophy: progress and potential. Mol. Diagn. Ther. *12*, 99-108.

Seoane,J., Le,H.V., and Massague,J. (2002). Myc suppression of the p21(Cip1) Cdk inhibitor influences the outcome of the p53 response to DNA damage. Nature *419*, 729-734.

Serrano,A.L., Murgia,M., Pallafacchina,G., Calabria,E., Coniglio,P., Lomo,T., and Schiaffino,S. (2001). Calcineurin controls nerve activity-dependent specification of slow skeletal muscle fibers but not muscle growth. Proc. Natl. Acad. Sci. U. S. A *98*, 13108-13113.

Shao,H., Chen,B., and Tao,M. (2009). Skeletal myogenesis by human primordial germ cell-derived progenitors. Biochem. Biophys. Res. Commun. *378*, 750-754.

Sharma,K.R., Mynhier,M.A., and Miller,R.G. (1993). Cyclosporine increases muscular force generation in Duchenne muscular dystrophy. Neurology *43*, 527-532.

Shaulian,E., Schreiber,M., Piu,F., Beeche,M., Wagner,E.F., and Karin,M. (2000). The mammalian UV response: c-Jun induction is required for exit from p53-imposed growth arrest. Cell *103*, 897-907.

Sifringer,M., Uhlenberg,B., Lammel,S., Hanke,R., Neumann,B., von,M.A., Koch,I., and Speer,A. (2004). Identification of transcripts from a subtraction library which might be responsible for the mild phenotype in an intrafamilially variable course of Duchenne muscular dystrophy. Hum. Genet. *114*, 149-156.

Skuk,D. *et al.* (2007). First test of a "high-density injection" protocol for myogenic cell transplantation throughout large volumes of muscles in a Duchenne muscular dystrophy patient: eighteen months follow-up. Neuromuscul. Disord. *17*, 38-46.

Speer,A. and Oexle,K. (2000). Muskeldystrophien. (in German). In: Handbuch der Molekularen Medizin (1.ed) Band 6 - Monogen bedingte Erbkrankheiten I, ed. D.Ganten and K.Ruckpaul; Berlin, Heidelberg: Springer Verlag, 3-30.

Spencer,M.J. and Mellgren,R.L. (2002). Overexpression of a calpastatin transgene in mdx muscle reduces dystrophic pathology. Hum. Mol. Genet. *11*, 2645-2655.

St-Pierre,S.J., Chakkalakal,J.V., Kolodziejczyk,S.M., Knudson,J.C., Jasmin,B.J., and Megeney,L.A. (2004). Glucocorticoid treatment alleviates dystrophic myofiber pathology by activation of the calcineurin/NF-AT pathway. FASEB J. *18*, 1937-1939.

Starke,P.H. (1990). Analyse Von Petri-Netz-Modellen (in German), München: Teubner.

Starke,P.H. and Roch,C. (1999). INA -The Integrated Net Analyzer. http://www2.informatik.hu-berlin.de/~starke/ina.html. Humboldt Universitaet, Berlin.

Steen,M.S., Adams,M.E., Tesch,Y., and Froehner,S.C. (2009). Amelioration of muscular dystrophy by transgenic expression of Niemann-Pick C1. Mol. Biol. Cell. *20*, 146-152.

Stupka,N., Gregorevic,P., Plant,D.R., and Lynch,G.S. (2004). The calcineurin signal transduction pathway is essential for successful muscle regeneration in mdx dystrophic mice. Acta Neuropathol. (Berl) *107*, 299-310.

Stupka,N., Schertzer,J.D., Bassel-Duby,R., Olson,E.N., and Lynch,G.S. (2008). Stimulation of calcineurin-A{alpha} activity attenuates muscle pathophysiology in mdx dystrophic mice. Am. J. Physiol Regul. Integr. Comp Physiol *294*, R983-R992.

Sun,L. *et al.* (2005). Calcineurin regulates bone formation by the osteoblast. Proc. Natl. Acad. Sci U. S. A *102*, 17130-17135.

Swat,M., Kel,A., and Herzel,H. (2004). Bifurcation analysis of the regulatory modules of the mammalian G1/S transition. Bioinformatics. *20*, 1506-1511.

Takeshima,Y., Nishio,H., Sakamoto,H., Nakamura,H., and Matsuo,M. (1995). Modulation of in vitro splicing of the upstream intron by modifying an intra-exon sequence which is deleted from the dystrophin gene in dystrophin Kobe. J. Clin. Invest *95*, 515-520.

Tidball,J.G. and Spencer,M.J. (2000). Calpains and muscular dystrophies. Int. J. Biochem. Cell Biol. *32*, 1-5.

Tidball,J.G. and Wehling-Henricks,M. (2004). Evolving therapeutic strategies for Duchenne muscular dystrophy: targeting downstream events. Pediatr. Res. *56*, 831-841.

Trimarco,A., Torella,A., Piluso,G., Ventriglia,V.M., Politano,L., and Nigro,V. (2008). Log-PCR: A New Tool for Immediate and Cost-Effective Diagnosis of up to 85% of Dystrophin Gene Mutations. Clin. Chem. *54*, 973-981.

Vainzof,M., yub-Guerrieri,D., Onofre,P.C., Martins,P.C., Lopes,V.F., Zilberztajn,D., Maia,L.S., Sell,K., and Yamamoto,L.U. (2008). Animal models for genetic neuromuscular diseases. J. Mol. Neurosci. *34*, 241-248.

van Deutekom,J.C. *et al.* (2007). Local dystrophin restoration with antisense oligonucleotide PRO051. N. Engl. J. Med. *357*, 2677-2686.

van Deutekom,J.C. and van Ommen,G.J. (2006). Novel RNA-based therapeutic PRO051 to enter Phase I/II clinical trials in DMD. Prosena in collaboration with the department of Human and Clinical Genetics. Leiden University Medical Center. The Netherlands. *http://prosensa.eu/*..

van Vliet,L., De Winter,C.L., van Deutekom,J.C., van Ommen,G.J., and artsma-Rus,A. (2008). Assessment of the feasibility of exon 45-55 multiexon skipping for Duchenne muscular dystrophy. BMC. Med. Genet. *9:105.*, 105.

Vieira,N.M., Brandalise,V., Zucconi,E., Jazedje,T., Secco,M., Nunes,V.A., Strauss,B.E., Vainzof,M., and Zatz,M. (2008). Human multipotent adipose derived stem cells restore dystrophin expression of Duchenne skeletal muscle cells in vitro. Biol. Cell *100*, 231-241.

Visser,D. and Heijnen,J.J. (2002). The mathematics of metabolic control analysis revisited. Metab Eng *4*, 114-123.

von Knebel,D.M., Bauknecht,T., Bartsch,D., and zur,H.H. (1991). Influence of chromosomal integration on glucocorticoid-regulated transcription of growth-stimulating papillomavirus genes E6 and E7 in cervical carcinoma cells. Proc. Natl. Acad. Sci. U. S. A *88*, 1411-1415.

von Moers,A. (2007). A combination of deflazacort and CsA may lead to improved conditions of DMD patients. Personal communication. Neuropaediatrics, Westend Klinikum, Berlin, Germany.

von Moers,A., Zwirner,A., Reinhold,A., Bruckmann,O., van,L.F., Stoltenburg-Didinger,G., Schuppan,D., Herbst,H., and Schuelke,M. (2005). Increased mRNA expression of tissue inhibitors of metalloproteinase-1 and -2 in Duchenne muscular dystrophy. Acta Neuropathol. (Berl) *109*, 285-293.

Walsh,F.S. (1990). N-CAM is a target cell surface antigen for the purification of muscle cells for myoblast transfer therapy. Adv. Exp. Med. Biol. *280*, 41-45.

Walsh,F.S., Parekh,R.B., Moore,S.E., Dickson,G., Barton,C.H., Gower,H.J., Dwek,R.A., and Rademacher,T.W. (1989). Tissue specific O-linked glycosylation of the neural cell adhesion molecule (N-CAM). Development *105*, 803-811.

Webster,C., Pavlath,G.K., Parks,D.R., Walsh,F.S., and Blau,H.M. (1988). Isolation of human myoblasts with the fluorescence-activated cell sorter. Exp. Cell Res. *174*, 252-265.

Weir,A.P., Morgan,J.E., and Davies,K.E. (2004). A-utrophin up-regulation in mdx skeletal muscle is independent of regeneration. Neuromuscul. Disord. *14*, 19-23.

Welch,E.M. *et al.* (2007). PTC124 targets genetic disorders caused by nonsense mutations. Nature *447*, 87-91.

Wells,D.J., Ferrer,A., and Wells,K.E. (2002). Immunological hurdles in the path to gene therapy for Duchenne muscular dystrophy. Expert. Rev. Mol. Med. *2002*, 1-23.

Wildermuth,M.C. (2000). Metabolic control analysis: biological applications and insights. Genome Biol. *1*, REVIEWS1031.

Wilton,S. (2007). PTC124, nonsense mutations and Duchenne muscular dystrophy. Neuromuscul. Disord. *17*, 719-720.

Winnard,A.V. *et al.* (1993). Characterization of translational frame exception patients in Duchenne/Becker muscular dystrophy. Hum. Mol. Genet. *2*, 737-744.

Wu,H.Y., Tomizawa,K., and Matsui,H. (2007). Calpain-calcineurin signaling in the pathogenesis of calcium-dependent disorder. Acta Med. Okayama *61*, 123-137.

Yoshida,H. *et al.* (1998). The transcription factor NF-ATc1 regulates lymphocyte proliferation and Th2 cytokine production. Immunity. *8*, 115-124.

Yu,Z. and Shah,D.M. (2004). U-937 monocyte-mediated c-Jun dephosphorylation and AP-1 activation in human endometrial stromal cells. Eur. J. Obstet. Gynecol. Reprod. Biol. *116*, 226-232.

Zhang,J., Gray,J., Wu,L., Leone,G., Rowan,S., Cepko,C.L., Zhu,X., Craft,C.M., and Dyer,M.A. (2004). Rb regulates proliferation and rod photoreceptor development in the mouse retina. Nat. Genet. *36*, 351-360.

Zhang,S.Z., Xie,H.Q., Xu,Y., Li,X.Q., Wei,R.Q., Zhi,W., Deng,L., Qiu,L., and Yang,Z.M. (2008). Regulation of cell proliferation by fast Myosin light chain 1 in myoblasts derived from extraocular muscle, diaphragm and gastrocnemius. Exp. Biol. Med. (Maywood.). *233*, 1374-1384.

Zhou,G.Q., Xie,H.Q., Zhang,S.Z., and Yang,Z.M. (2006). Current understanding of dystrophin-related muscular dystrophy and therapeutic challenges ahead. Chin Med. J. (Engl) *119*, 1381-1391.

9 Appendices

9.1 Glossary

Boolean networks	A network that consists of Boolean variables the state of which is defined by other network variables. Boolean is a computational logic that sets data value to true or false (usually 1 and 0, respectively).
Tanimoto coefficient	A normalised simple distance measure, in this case based on the Parikh vector is used as distance measure. Vectors t_i and t_j describing two t-invariants is defined for *Parikh vector* and support as: $$s(t_i, t_j) = s_{ij} = \frac{|(t_i)\,(t_j)|}{|(t_i)\,(t_j)|}$$ where (t_i) and (t_j) denote vectors of the t-invariants. The pair-wise similarity s_{ij} expressed by this coefficient is transformed into a distance d_{ij} by $d_{ij} = 1 - s_{ij}$.
Firing rule	A transition t 0 T can fire in a marking, m, if t is enabled in m. After firing of t, a successive marking, m', is reached with $m' := m + \Delta t$. The minimum number of tokens are defined on the pre-places by the marking $t^-(p) := W(p, t)$, if (p, t) 0 F and $t^-(p) := 0$, if (p, t) ∫ F, and the number of tokens, which are added to each post-place by $t^+ := W(t, p)$, if (t, p) 0 F and $t^+(p) := 0$, if (t, p) ∫ F. $\Delta t := t^- - t^+$ gives the change of marking in the considered place.
Incidence matrix	A $(k \times l)$- matrix, c, with k as number of places and l as number of transitions. Every matrix element $c_{i,j}$ corresponds to the token change on place p_i by firing of transition t_j.
Liveness of Petri nets	A Petri net is live if every transition within the net is live, and is dead if all transitions are dead in the initial marking. A dead transition is not in any firing sequence. Petri net $N = (P, T, F, W, m_0)$, with m an arbitrary marking in N, and t 0 T an arbitrary transition. • A transition t is *live* in the marking, m in N, if for every marking, m' 0 $R_N(m)$, a further marking m'' 0 $R_N(m')$ with $m' \overset{t}{\rightarrow} m''$ exists. • A transition t is *dead* in the marking, m in N, if for every marking, m' 0 $R_N(m)$, holds: $t^- \Leftarrow m'$. • A marking, m, is called *live* in N, if all transitions, t 0 T, are *live* in m. • A marking, m, is called *dead* in N, if all transitions, t 0 T, are *dead* in m. • A transition t is called *live (dead)* in N, if t is live *(dead)* in m_0. • A Petri net, N, is called *live (dead)*, if m_0 is live *(dead)* in N. • A Petri net, N, is called *deadlock-free*, if there is no reachable marking, where all transitions, t 0 T in N, are *dead*.

APPENDICES

Mauritius maps	Petri net $N = (P, T, F, W, m_0)$; X is denoted as the set of all n T-invariants x. A finite binary tree, $T = (V, E)$, is called Mauritius map, if • the set V is a finite set of transitions belonging to a t-invariant, x. The root vertex is located in the lower left corner. • the set $E = (H, R)$ is a finite set of edges between vertices, indicating dependencies of t-invariants. • the set H represents horizontal edges, which connect vertices belonging to the same t-invariant. • the set R represents vertical edges, which connect vertices of the left subtree with vertices of the right upper subtree belonging to the same t-invariant.
MCT-sets	X is denoted as the set of all n T-invariants x. Two transitions t_i and t_j belong to the same MCT-set, if and only if they participate in exactly the same T-invariants: $\forall i, j\ 0\ prot$, t_i and t_j correspond to the same MCT-set, if and only if $\forall x\ 0\ X, t_i\ 0\ supp(x), t_j\ 0\ supp(x)$. This dependency relation leads to maximal common sets of transitions. A transition set $A \subseteq T$ is called an MCT-set, if and only if $\forall x\ 0\ X : A \subseteq supp(x)\ \omega\ A\ 1\ supp(x) = \emptyset$.
Minimal invariant	An invariant u is called minimal if its support supp (u) does not contain the support of any other invariant supp (z), i.e. $\int invariant\ z : supp\ (z)\ \delta\ supp\ (u)$, and the largest common divisor of all non-zero entries of u is equal to one. In the following we refer to minimal invariants writing invariants for short. Trivial t-invariants consist of two transitions describing the forward or backward reaction of a reversible reaction.
ODE	Ordinary Differential Equation
P-invariant	Vector definition $y \geq 0$, $y\ 0\ N^k_0$ that verifies $C^T \cdot y = 0$. It stands for a set of places, over which the weighted sum of tokens is always constant. For a P-invariant y and any markings m_1, $m_2\ 0\ N^k_0$ that are reachable by the firing of transitions, it holds $y \cdot m_1 = y \cdot m_2$.
Petri nets	Bipartite directed labbled-graphs consisting of two types of nodes: places and transitions with (P, T, F, W, m_0), if it holds: • P and T are two sets with $P \cap T = \emptyset$, $P \perp T \neq \emptyset$ The elements of the sets P and T are called places and transitions, respectively. • F is a two digit relation with $F \subseteq (P \times T) \perp (T \times P)$. The elements of F are called arcs. F is called the flux relation of N. • $W : F \to N$, is the weight of the arcs. • $m_0 : P \to N_0$, is the initial marking of N. The directions of the edges define for each transition a set of pre-places and a set of post-places, denoted as •t := ch and t • := {p	p 0 P ϖ (t, p) 0 F }, respectively. Accordingly, there are for each place a set of pre-transitions and a set of post-transitions, denoted as •p := {t	t 0 T ϖ (t, p) 0 F } and p • := {t	t 0 T ϖ (p, t) 0 F }, respectively.

Places	Passive system elements; e.g. chemical compounds, proteins complexes, genes. $P = \{p_1, \ldots, p_l\}$
Reachability graph	Petri net $N = (P, T, F, W, m_0)$: A graph is called a reachability graph of N, if: • The set R_N represents the vertices of the graph. • (m, m') denotes an edge of the graph, if a transition t exists with $m' \xrightarrow{t} m''$.
Support of a vector	The set of non-zero elements of a vector is called the support of the vector u, written as *supp (u)*.
t-invariant	Describes the system behaviour of the network. Vector definition based on the incidence matrix: vector $x \geq 0$, x 0 N^l_0 with the equation $C \cdot x = 0$.
Tokens	Represent a number of molecules or chemical compounds
Transitions	Active system elements; e.g. chemical reactions. $T = \{t_1, \ldots, t_l\}$

9.2 Publication list

Original-Article

Sifringer M, Uhlenberg B, **Lammel S**, Hanke R, Neumann B, von Moers A, Koch I, Speer A. Identification of transcripts from a subtraction library which might be responsible for the mild phenotype in an intrafamilially variable course of Duchenne muscular dystrophy. Hum Genet. 2004 Jan;114(2):149-56.

Luhmann UF, Lin J, Acar N, **Lammel S**, Feil S, Grimm C, Seeliger MW, Hammes HP, Berger W. Role of the Norrie disease pseudoglioma gene in sprouting angiogenesis during development of the retinal vasculature. Invest Ophthalmol Vis Sci. 2005 Sep;46(9):3372-82.

Grunwald S and Speer A. Efficient and minimal toxic transfection of primary human skeletal myoblasts using Fugene HD transfection reagent. Roche Biochemica; Juli 07; 3: 26-27

Grunwald S, Speer A, Ackermann J, Koch I. Petri net modelling of gene regulation of the Duchenne muscular dystrophy. Biosystems; 2008 May;92(2):189-205

Grafahrend-Belau E, Schreiber F, Heiner M, Sackmann A, Junker BH, **Grunwald S**, Speer A, Winder K, Koch I. Modularization of biochemical networks based on classification of Petri net t-invariants. BMC Bioinformatics. 2008 Feb 8;9:90.

Conferences/ Workshops

Poster

Congress – European Society of Gene and Cell Therapy (ESGCT) Prag 2005
Grunwald S, von Moers A., Koch I, Hobbiebrunken E, Wilichowski E, Speer A. Identification and characterisation of phenotype modifying genes and pathways in Duchenne muscular dystrophy patients: suggestions for a dystrophin downstream gene therapy.

Congress – German Society of Muscle Diseased Persons (DGM) Freiburg 2007

von Moers A, **Grunwald S***, Koch I, Kamp R, Lochmüller H, Speer A. Identifizierung und Charakterisierung von Muskeldystrophie Duchenne modifizierenden Genen und Stoffwechselwegen zur Entwicklung neuer Therapiestrategien.

* gleichberechtigter Erstautor

Symposium – Berlin Center for Genome based Bioinformatics (BCB) Berlin 2007

Grunwald S, Speer A, Ackermann J, Koch I. Petri net modelling of gene regulation in Duchenne muscular dystrophy.

Conference – Intelligent Systems for Molecular Biology (ISMB) Wien 2007

Grunwald S, Speer A, Ackermann J, Koch I. Analysis of gene regulation in Duchenne Muscular Dystrophy by Petri net modelling.

Congress – European Society of Gene and Cell Therapy (ESGCT) Brügge 2008

Grunwald S, von Moers A., Koch I, Speer A. The combined application of deflazacort and cyclosporin A enhances utrophin expression in skeletal muscle cells of DMD patients in cell culture

Vorträge (* speaker)

Conference – Computational Molecular Biology Moskau 2007

Koch I*, **Grunwald S**, Ackermann J, Speer A. Modelling and analysis of molecular processes in Duchenne Muscular Dystrophy using Petri nets.

Workshop –Data and Knowledge Based Biomolecular Network Reconstruction Jena 2007

Koch I*, **Grunwald S**, Ackermann J, Speer A. Modelling biochemical processes of Duchenne Muscular Dystrophy with Petri nets.

Workshop – Molecular Interactions Berlin 2008

Koch I*, Ackermann J, **Grunwald S**, Speer A

Modelling of pathomechanism of Duchenne Muscular Dystrophy.

Die VDM Verlagsservicegesellschaft sucht für wissenschaftliche Verlage abgeschlossene und herausragende

Dissertationen, Habilitationen, Diplomarbeiten, Master Theses, Magisterarbeiten usw.

für die kostenlose Publikation als Fachbuch.

Sie verfügen über eine Arbeit, die hohen inhaltlichen und formalen Ansprüchen genügt, und haben Interesse an einer honorarvergüteten Publikation?

Dann senden Sie bitte erste Informationen über sich und Ihre Arbeit per Email an *info@vdm-vsg.de*.

Sie erhalten kurzfristig unser Feedback!

VDM Verlagsservicegesellschaft mbH
Dudweiler Landstr. 99
D - 66123 Saarbrücken

Telefon +49 681 3720 174
Fax +49 681 3720 1749

www.vdm-vsg.de

Die VDM Verlagsservicegesellschaft mbH vertritt

Printed by Books on Demand GmbH, Norderstedt / Germany